MANAGING HEALTH PROMOTION IN THE WORKPLACE

COMMITTEE TO DEVELOP GUIDELINES FOR HEALTH PROMOTION PROGRAMS AT THE WORKSITE

Committee Members

REBECCA S. PARKINSON, M.S.P.H., Chairperson
Staff Manager, Employee Health Education
Corporate Medical Division
American Telephone & Telegraph Company

ROBERT N. BECK, M.S.
IBM Director, Benefits and Personnel Services
International Business Machines Corporation

G. H. COLLINGS, JR., M.D., M.P.H.
Corporate Medical Director
New York Telephone Company

MICHAEL ERIKSEN, SC.M.
Community Health Educator
Health Education Center
Maryland Department of Health and Mental Hygiene

ALICE M. MCGILL, PH.D.
Office of Health Information, Health Promotion and Physical Fitness and Sports Medicine
Department of Health and Human Services

CLARENCE E. PEARSON, M.P.H.
Assistant Vice President
Health and Safety Education Division
Metropolitan Life Insurance Company

BEVERLY G. WARE, DR.P.H.
Corporate Health Education Programs Coordinator
Ford Motor Company

Cosponsoring Agencies (Ex Officio)

NATIONAL CENTER FOR HEALTH EDUCATION

Donald Merwin
Vice President for Program Development

OFFICE OF HEALTH INFORMATION, HEALTH PROMOTION AND PHYSICAL FITNESS AND SPORTS MEDICINE (DHHS)

Lawrence W. Green, Dr.P.H.
Former Director, Office of Health Information and Health Promotion
Visiting Lecturer, Harvard Medical School
Director, Center for Health Promotion Research and Development, University of Texas Health Science Center

MANAGING HEALTH PROMOTION IN THE WORKPLACE

Guidelines for Implementation and Evaluation

Rebecca S. Parkinson and Associates

Mayfield Publishing Company

First edition 1982

Library of Congress Catalog Card Number: 81-84693
International Standard Book Number: 0-87484-567-X

Manufactured in the United States of America
Mayfield Publishing Company
285 Hamilton Avenue
Palo Alto, California 94301

Sponsoring editor: C. Lansing Hays
Manuscript editor: Liese Hofmann
Designer: Marie Carluccio
Production Manager: Cathy Willkie
Compositor: Imperial Litho/Graphics
Printer and binder: Bookcrafters

Contents

III BACKGROUND PAPERS

Foreword

Corporate interest in health promotion is very much in evidence today, as exhibited by the steady growth of employee health promotion programs nationwide. Many business leaders have come to realize that an investment in health promotion and disease prevention holds the prospect of improved employee productivity and substantial long-term cost savings.

Total medical care expenditures in the United States in 1960 were $26.9 billion (5.3 percent of the Gross National Product). By 1970 those costs had risen to $75 billion (7.6 percent of the GNP), and by 1980 the costs were recorded at $243.4 billion (9.4 percent of the GNP), with business paying over half of the national health care bill. If permitted to continue, health costs or, more accurately, the costs of illness and rehabilitative care, are expected to top $462.2 billion in 1985 (9.9 percent of the GNP), reflecting yearly per capita expenditures of $1,946.50.

As a result, businesses have been compelled to take steps enlisting health promotion efforts that offer the potential for slowing this trend and reducing the costs associated with absenteeism, hospitalization, disability, job turnover, and premature death. Corporations are not limiting their concern to the health and well-being of their employees. Because the health care costs incurred by a company come also from the family and the retiree—not just from the active employee—many worksite health promotion programs are now being broadened to include dependents and retirees.

The worksite is one of the most promising and challenging sites for health behavior change. It is, after all, where an individual spends nearly 30 percent of his or her waking hours. Health promotion programs offered on company grounds or nearby in the community facilitate the involvement of persons who are on tight schedules owing to family and social responsibilities. In addition, the worksite offers accessibility to large groups of people and a ready conduit through which their families can be reached. Thus, health promotion programs extend far beyond the worksite to influence the health and well-being of the community.

A number of efforts to facilitate the growth of employee health promotion programs have been sponsored in both the private and the public sectors. The National Center for Health Education (NCHE) has had a long-standing interest in occupational health promotion. And the federal Office of Health Information and Health Promotion (OHP) in 1979 sponsored the first broad national conference on Health Promotion Programs in Occupational Settings. With representation from business, labor, health, and academic institutions, that conference and its proceedings helped both to stimulate further interest in these programs and to highlight the need for materials to assist their implementation on a broader scale.

The guidelines in this book have been developed to advance these efforts. Through a committee established as a joint public/private venture by the OHP and the NCHE, a carefully structured process has been under way to provide general guidance for organizations interested in developing employee-based health promotion activities. The guidelines cover a broad range of topics: assessment of needs, specification of program objectives, organizational placement, approaches to implementation, identification and allocation of resources, and evaluation. A number of professional and voluntary organizations, companies, and individuals from both the public and the private sectors have offered helpful comment on the guidelines. The role of the OHP and the NCHE has been principally that of a catalyst, and the volume that has emerged represents the combined opinions of those who represent health promotion in industry today.

We congratulate the committee on this product and acknowledge the contributions of each member to the contents. Special thanks go to Dr. Lawrence Green, former director of the OHP, to Dr. Alice McGill of the OHP, to Mr. Donald Merwin of the NCHE, and of course to Ms. Rebecca Parkinson, who chaired the committee and directed the editorial process. We feel they have produced a valuable resource in the promotion of better health for Americans.

J. Michael McGinnis, M.D.
Deputy Assistant Secretary for Health
(Disease Prevention and Health Promotion)
U.S. Department of Health and Human Services

Merlin K. DuVal, M.D.
President
National Center for Health Education

Executive Summary

Interest in disease prevention, better health, self-care, and self-help activities has been gaining momentum among industry and union leaders, government officials, health care professionals, and the American public. This interest, together with soaring medical costs and the imperative of OSHA regulations, has focused industry concern on health promotion programs. Employees too have begun to take initiatives in this area.

What does health promotion at the worksite mean? In this context health promotion is a combination of educational, organizational, and environmental activities designed to support behavior conducive to the health of employees and their families. First an assessment is made of the risks of major sickness, disease, and premature death among employees. Interventions are then introduced to reduce such risks associated with behavior.

Is there evidence that health promotion programs work? Yes. The major health concerns of employees and their families are well known. Indicators of risk have been identified and quantified. A substantial majority of chronic diseases are known to be linked with life-style, and certain educational, organizational, and environmental interventions have already proved effective.

Is there evidence that health promotion programs are cost-effective? There is as yet no hard evidence. But potentially significant costs could be reduced, such as insurance premiums, disability benefits, and medical expenses. Involving the employees' families in the programs is clearly prudent, because about two-thirds of these costs are incurred by dependents. Moreover, since the target population is a fairly large, well-defined group of people, generally accessible to educational or organizational change and influential in the larger community, the workplace is in many respects an ideal setting for a health promotion program. (Admittedly, "ideal" applies more to white-collar businesses than to heavy industry or to occupations that are hazardous by nature.)

This book is intended primarily for managers of businesses and unions and for health professionals in such organizations. Its aim is to pave the way as they explore how to set up a health promotion program or upgrade an existing one. The focus is on identifying the risks of major disease and premature death among employees and their families and on intervening to minimize those risks by supporting healthy behavior changes with organizational and environmental arrangements. Although this book does not, except in one background paper, cover occupational hazards like exposure to toxic substances, the proposed programs are meant to complement traditional and necessary health and safety measures.

DESIGNING THE PROGRAM

Five elements are essential to an effective health promotion program:

1. *Assessment of needs.* Objectives are best set after needs have been clearly defined. If other health promotion activities are currently offered, how effective are they? Do employees perceive a need for the proposed program? What are their sociodemographic characteristics? What are their health habits? What is the prevalence of risk among the employees, and what are the clinical indicators of risk? What are your data on the utility and cost of medical services; costs of disability benefits and insurance premiums; and occurrence and costs of disability and incidental absenteeism?

2. *Setting priorities and objectives.* Since health is relative and not always easily quantified, a program's goals, based on a needs assessment, are ordinarily stated and measured either as (1) the cumulative benefits of better health (e.g., fewer absences, less frequent and less severe disease) or (2) intermediate results (e.g., greater awareness of and knowledge about health matters, improved behavior, reduced risk). The first category of objectives requires longer-term resource investment than does the second category.

3. *Organizational location.* The medical, benefits, and personnel departments are the most common locations of the administration of these programs. Each has its pluses and minuses.

4. *Implementation strategies.* Key matters are the selection of the most appropriate organizational location for your program; allocation of resources; the type(s) of educational approach to be used; timing, publicity; incentives; and even some ethical issues. The program can be

approached on a pilot basis, phased in, or implemented totally at the outset. Each method has its advantages and disadvantages.

5. *Identification and allocation of resources.* What personnel will be needed? Will there be outside personnel? Will any personnel need special instructions? Are physical facilities and educational materials adequate? No single educational approach or behavioral intervention provides the final answer in supporting changes in health behavior; a combination of approaches is ordinarily more effective. One approach could include large group meetings to convey information, heighten awareness, and provide opportunities for questions and answers; another could include small group meetings, or person-to-person discussions, to give employees individual attention, social support, and a chance to talk over their particular concerns. Whatever your approach, due consideration must be given to incentives, publicity, funding, and the site and time of the activities.

EVALUATION

Evaluation is essential. Is the program actually improving your employees' health? Is it cost-effective?

The evaluation can focus on the process (the manner in which program activities are being carried out) as well as on the immediate effects and ultimate outcomes of the program. Short-term effects might include improved knowledge of cardiovascular risk factors; greater belief in the preventability of chronic disease; increased levels of regular aerobic exercise; less cigarette smoking; lower consumption of high-cholesterol foods; better adherence to regimens to control high blood pressure. Long-term effects might include increased productivity; less absenteeism; fewer health care claims; reduced cardiovascular morbidity and mortality.

The success of your program can be gauged either by change from a previous situation by comparison with a group of employees not receiving the program or by a progress toward, or achievement of, some predetermined goal. Depending on the needs and resources of your organization, the evaluation process can range from a simple historical, recordkeeping approach, to more complex periodic inventories, to controlled experiments, to a full-scale scientific research project.

CONCLUSIONS

The planning of a health promotion program is similar to the planning of any major business venture. It requires an assessment of needs, the setting of priorities and objectives on the basis of those needs, the placement of administrative responsibility, the design or selection of strategies, the allocation of resources, and evaluation. The most common error in some of the early programs was their oversimplification of these steps. A systems approach, including evaluation, will provide feedback to the organization, assuring the continuous improvement of the program.

Introduction

Increasing consumer interest in disease prevention, better health, self-care, and self-help activities has prompted industry leaders, union officials, government policymakers, and third-party payers to consider seriously the merits of health promotion programs at the worksite.

In 1979, when the nation's health expenditures were over $212 billion, 92.5 percent of those questioned in a Harris poll agreed with this statement: "If we Americans lived healthier lives, ate more nutritious food, smoked less, maintained proper weight and exercised regularly, it would do more to improve our health than anything doctors and medicine could do for us." But there are all those unsettling statistics that reveal the wide discrepancy between belief and behavior. For instance, nearly half of Americans are overweight by some criteria and fewer than a third follow the widely publicized recommendation to exercise vigorously at least three times a week.

Nevertheless, it is encouraging that heightened consumer interest in health matters, along with better medical care, has already led to some notable improvements. For example, although cardiovascular disease remains the nation's leading cause of death, the mortality rates for this disease have dropped more than 35 percent since 1950, and over two-thirds of this decline occurred in the last decade. Further, in that decade, both men and women gained more than two years in life expectancy. These improvements have been attributed not only to primary prevention and better medical treatment but, importantly, to changes in risk factors: for instance, less consumption of high-cholesterol foods, less cigarette smoking, more regular exercise, and greater awareness of and more aggressive treatment of hypertension *(1)*.

Clearly, people have a choice. They can, through ignorance or apathy, perpetuate poor health habits that continue to exact a high toll in chronic disease and premature death, or they can make changes that will improve their health, no matter how bad it may be, and reduce risk factors. But people often need motivation and help in learning how to take care of themselves. Here is a unique opportunity for employers to respond with effective health promotion programs for their employees.

The time is right. On the national level, federal interest in health promotion is reflected in the 1978 Report of the HEW Departmental Task Force on Prevention *(2)*, the 1979 Surgeon General's Report, *Healthy People* *(3)*, the 1979 National Conference on Health Promotion Programs in Occupational Settings *(4)*, and the 1980 Objectives for the Nation *(5)*—all designating disease prevention and health promotion as major health initiatives. Skyrocketing health-related costs, coupled with the imperative of OSHA regulations, have further spurred employers to initiate efforts to reduce the risk of disease, disability, and premature death among their employees. The goal is to decrease health care costs and increase productivity as well as provide personal benefit to the employee and his or her family through a healthier and possibly longer life.*

As companies and unions contemplate instituting health promotion programs to complement their existing health and safety programs, three common questions emerge:

1. What is the nature of health promotion at the worksite?
2. What evidence is there that such programs will work?
3. Is there evidence that money expended for these programs will yield an equivalent or greater return?

WHAT IS HEALTH PROMOTION AT THE WORKSITE?

Health promotion is a combination of educational, organizational, and environmental activities designed to support behavior conducive to the health of employees and their families.

The most effective health promotion programs include the following components:

- *Assessment of risk.* Through a health risk appraisal or other measure, individual risk of morbidity or premature mortality is assessed for car-

---

*Throughout this book the term "employee" is meant to refer to both employees and retirees.

diovascular disease, cancer, stroke,* mental health problems, and accidents.

♦ *Risk reduction.* Ideally, all the following would make up a health promotion program, but one or more might be selected, depending on your organization's needs and constraints:

—High blood pressure control

—Smoking control

—Drug/alcohol abuse control

—Weight control and nutrition education

—Exercise/physical fitness

—Early cancer detection

—Accident prevention/self-protective measures against environmental and other health hazards at the workplace

—Stress management

♦ *Evaluation.* This is essential to determine whether the program has led to:

—Reduction in risk of cardiovascular disease, cancer, stroke, mental health problems, and accidents

—Decrease in absenteeism

—Increase in morale and in productivity

—Decrease in medical costs, insurance premiums, and disability benefits

—Reduction in morbidity and premature mortality

♦ *Environmental and social support.* The environment—physical, social, political, and economic—that supports life-style change is an important element in a program. The program should therefore include some or all of the following:

—Employee participation in teaching and program management

—Employee support groups

—Cafeteria programs for more nutritious foods

—Vending machines with nutritious snacks

---

*Because a stroke is a cerebrovascular accident resulting in a neurological disorder, in these guidelines it is not termed, as is commonly done, a cardiovascular disease, even though it usually stems from cardiovascular disease.

—Outdoor recreation facilities
—Ergonomic considerations
—Lending library of health books and journals

WHAT EVIDENCE IS THERE THAT SUCH PROGRAMS WILL WORK?

The pivotal question is whether enough is known about instituting these programs to proceed with their implementation or whether employers should wait until more evidence is available. At this time the following is known:

- The major health concerns of this country are cardiovascular disease, cancer, stroke, mental health problems, and accidents.
- Risk indicators—like smoking, hypertension, elevated serum cholesterol levels, excessive alcohol consumption, and obesity—can be both identified and quantified.
- A substantial majority of chronic diseases are linked with life-style.
- Certain educational, organizational, and environmental interventions have proved effective in assisting persons to change their life-style and thus prevent or control chronic disease.

ARE THE PROGRAM BENEFITS GREATER THAN THE COSTS?

Evidence of the cost benefits of health promotion efforts, particularly at the worksite, is just beginning to grow. For instance, during the fiscal years 1973–74 and 1974–75, hypertension and its complications resulted in 23,129 days of absence among the 86,000 employees at the New York Telephone Company. The direct cost to the company has been estimated at $1,040,805, or about $500,000 a year. In the area where the company's health promotion program, called Health Care Management, was operating, there was a 43 percent reduction in the costs of absenteeism stemming from hypertension (6). The cost benefits of similar efforts to combat hypertension are discussed in Chapter 25.

Cancer prevention programs at the worksite also offer evidence of cost benefits. For example, among persons aged 45 to 65 years, one out of 1,000 is likely to get colo-rectal cancer, and about 40 percent die of the disease. In 10 years, then, there would probably be 10 colo-rectal cancers (each requiring surgery) and 4 deaths. The total cost of surgery (i.e., the sum of the costs for the surgeon, hospital, and sickness period) for 10 patients would be around $300,000. At the Campbell Soup Company (Camden, N.J.) a colo-rectal cancer prevention program (including sigmoidoscopies, polypectomies, and double-contrast X rays as needed) comes to $10,000 a year for 1,000 persons—or $100,000 in 10 years. Only one employee has died of colo-rectal cancer since the program began. The cost of treating this patient, a machinist, was $66,000 (over 4½ years), and this cost was in insurance payments only, with $14,000 in life insurance. Thus the preventive program appears to have reduced the incidence of colo-rectal cancer at a 10-year cost of $100,000 for 1,000 persons, in contrast to a higher death rate and a cost of $300,000 without the program. Obviously, if company-paid life insurance were more than $14,000 per person, potential savings would significantly increase *(7)*.

Some companies have already instituted health promotion programs despite lack of data on cost benefits. The potential payoff is the long-term possibility of reducing insurance premiums, disability benefits, and medical costs. Many companies are also interested in improving not only the health of their employees but that of the family members and retirees as well.

In many respects the workplace is an ideal setting for a health promotion program. Such a program complements existing health and safety activities. The worksite offers (1) access to a large number of people; (2) social support networks to assist employees in improving their life-style; (3) the opportunity for an organization's health professionals to become an integral part of the program; and (4) a focal point for raising the level of health of an entire community.

The guidelines offered in this book should help you deal with basic questions that arise in planning, implementing, and evaluating a health promotion program:

- How does one establish a health promotion program?
- What should the objectives of the program be?
- What target group(s) should the program serve?
- What kinds of personnel and physical resources are needed?
- In which department should the administration of the program be located?
- What are the alternatives for implementing the program?
- How can the program be evaluated for short- and long-term impact?

Although many existing health promotion programs have involved large companies making substantial financial commitments, this should not deter smaller companies from setting up such programs on a different scale.

It may be practical for several small companies jointly to finance a program that can be offered to all their employees. The experience of companies that have already implemented programs is a valuable resource that can be used by large and small companies alike (see Part II for examples of existing programs).

Healthy workers are an asset to themselves and their families, to management, to labor organizations, to stockholders, and to the community at large. As the experience of both the private and the public sector in implementing health promotion programs grows, so too will the cooperative effort to ensure a healthier life-style for working men and women and for their families as well.

REFERENCES

1. Levy, Robert I. The decline in cardiovascular disease mortality. *Annual Review of Public Health* 2:49–70 (1981).
2. U.S. Department of Health, Education, and Welfare. *Departmental task force on disease prevention and health promotion: Federal programs and prospects.* Washington, D.C.: Government Printing Office, 1978.
3. U.S. Department of Health, Education, and Welfare. Office of the Assistant Secretary for Health and Surgeon General. *Healthy people: The Surgeon General's report on health promotion and disease prevention.* Washington, D.C.: Government Printing Office, 1979.
4. Department of Health and Human Services. *Proceedings of the national conference on health promotion programs in occupational settings.* Washington, D.C.: Government Printing Office, 1979.
5. Department of Health and Human Services. *Promoting health/preventing disease: objectives for the nation.* Washington, D.C.: Government Printing Office, 1980.
6. Collings, G.H. New York Telephone multiphasic screening program. In *High blood pressure control in the work setting: Issues, models, resources.* Proceedings of the National Conference, Washington, D.C., October 14, 1976. West Point, Pa.: Merck, Sharpe and Dohme, n.d.
7. Wear, Roland F., Jr., M.D., Corporate Medical Director, Campbell Soup Company, Camden, N.J. Personal communication, July 13, 1981.

Procedures for developing a health promotion program at the worksite are described in detail in this part of the book. The basic tasks of identifying needs and setting objectives are covered in Chapter 1. Chapter 2 presents the essentials for designing the program: how to build objectives based on needs, where to locate the program in the organization, how to implement the proposed program, and how to identify and allocate resources. Chapter 3 rounds out these guidelines with a discussion of evaluation—the part of a program that too often is underrated and tends to be conducted haphazardly. Six designs for evaluation are offered, from the simple recordkeeping approach to the full-blown evaluative research project.

I

GUIDELINES

1

A Hierarchy of Program Needs and Objectives

IDENTIFYING NEEDS AND SETTING OBJECTIVES

In determining the direction and scope of a health promotion program, the preferred procedure involves three steps. First, identify the needs and risks related to health. Second, set priorities. Third, specify objectives. This chapter discusses all three steps. The first step will be presented in procedural detail in the next chapter.

Health is, of course, a relative matter, and is not always easily quantified. So the needs and objectives for a health promotion program are usually stated and measured either as (1) the cumulative benefits of better health (e.g., fewer absences, less frequent and less severe disease) or as (2) intermediate results (e.g., greater awareness of and knowledge about health matters, improved behavior, reduced risk).

Business managers usually prefer the first category because they find it difficult to relate the value of the intermediate results to their ever-present need to justify health promotion programs as being confluent with their primary business objectives. But the data required to measure cost, morbidity, and mortality outcomes frequently must be organized from outside sources. Moreover, such data often are not readily available in a usable form. Achieving the objectives that call for reductions in cost, morbidity, and mortality therefore requires longer-term resource investment. Consequently, from the health promotion point of view, intermediate results are more attractive as objectives for the short term. They are quantifiable by data that can be produced within the project itself and are generally more usable in the actual management of the program.

Which level of outcome measurement would be most valuable and practical for your company?* Your decision can be based on a continuum that orders the needs and desired results at increasing levels of complexity. These levels pertain not only to the outcome to be achieved (see Figure 1.1) and the methodologies to achieve it, but also to the means of assessing its merit (see Chapter 3).

LEVELS OF NEEDS AND OBJECTIVES

Awareness

Creating awareness is often the most immediate need and therefore the first objective of a health promotion program. The need and the intent are to help employees become aware of their health habits and of the specific consequences of the detrimental ones so that they can take action to modify their life-style accordingly. In addition, the intent may be to raise the level of awareness concerning environmental effects on health. Heightening awareness is the simplest level of need or objective. It may be accomplished with as little effort as the use of the media.

Awareness will be heightened most when:

- The intended audience is in general motivated to gain more understanding about health problems.
- The intended audience is an active rather than a passive recipient of the information.
- Communication is through several print/audiovisual methods.
- Communications do not create unrealistic expectations among employees, given the personnel and finances available.

Knowledge

When the need or objective is increased knowledge, the emphasis is on providing information. For example, information about types of exercise, nutrition, the effects of smoking, and signs of cancer can be offered. To ascertain positive change, pre- and post-tests of knowledge levels are usually obtained and educational methods are employed to acquaint employees with

*To avoid complex syntax, the word "company" is ordinarily used in these guidelines to indicate both management and labor organizations.

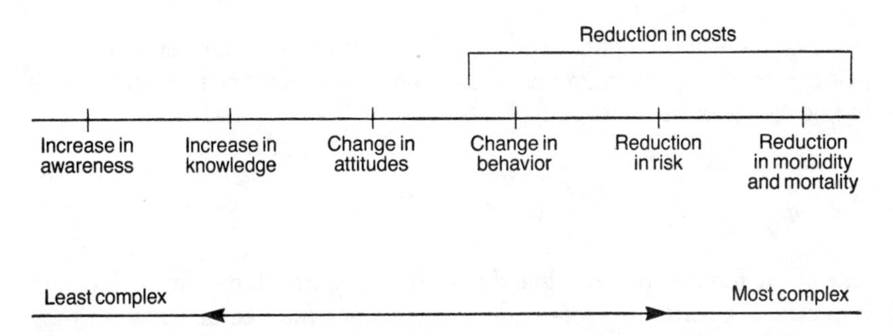

FIGURE 1.1 Levels of needs and objectives for health promotion programs
Items on this continuum are not always causally related to one another, nor do they necessarily represent a sequential series of events. It is generally true, however, that a program addressed to a more complex need or objective must include components that address the preceding, less complex objectives.

specific health-related information. An understanding of the information gained can be demonstrated by an ability to recognize, discuss, analyze, and synthesize verbally or in writing the information presented on a particular health topic.

Knowledge will be increased most when:

- Information is presented clearly and concisely.
- Materials used are geared to the audience's reading and general educational levels.
- The source of the information is perceived as credible by the intended audience.
- Scientific facts are documented.
- Issues presented are fully disclosed in an objective manner, especially when dealing with controversial subjects.

Attitudes and Beliefs

Attitudes and beliefs of employees are often major determinants of behavior affecting health. A change in attitude usually requires complex educational intervention and regular, long-term reinforcement. Attitude change is often associated with belief patterns and with the degree of internal control that persons perceive they have over their own health *(1, 2, 3)*. Recent research suggests that although changes in knowledge and attitudes are important, they often bear scant relation to actual changes in health behavior. More and more educational programs are focusing on behavior change as the end point, even though such programs may not have affected significant changes in knowledge or attitude.

Behavior

One of the fundamental needs and objectives of a health promotion program is to create opportunities for employees to adopt healthier habits by learning new behavioral patterns or skills.

When the program need or objective is behavior change, an initial assessment of the population's health habits is usually made. Educational opportunities and skill-building programs are then offered to employees to enable them to make the necessary changes in health habits. Periodic reassessments determine retention or change over time.

Although some behavior changes may occur within a relatively short time, their maintenance is much more difficult. Behavior is, after all, a function of health habits, personality characteristics, sociocultural influences, skills, attitudes, beliefs, values—the list is endless. As the complexity of the behavior change increases, so do the complexities of the educational approaches and evaluation methodologies needed to initiate and, especially, maintain the change.

Risk Reduction

It is now possible to quantify with reasonable precision a person's risk for cardiovascular disease, cancer, stroke, mental health problems, and accidents. Questionnaires, known as health risk appraisals, can be used to estimate risks from a number of activities or conditions—for instance, smoking, excess drinking, uncontrolled high blood pressure—that are associated with specific diseases, illnesses, disabilities, and premature death (see Chapter 22). Such questionnaires are often used to gather baseline information and to assess changes in risk periodically following an educational risk-reduction program. Three to five years are often necessary to assess clinically significant changes in risk. Meanwhile, shorter-term measures can be used to assess the progress of the program in achieving modification of behaviors associated with risk.

Reduction in Morbidity and Premature Mortality

Reduction in the morbidity and premature mortality that stem from cardiovascular disease, cancer, stroke, and other health problems is an ambitious goal that may take a long time to be realized. A series of steps involving risk reduction and behavior change, and their maintenance over time, are required before any changes in morbidity and mortality rates can be detected.

CONCLUSIONS

The needs and objectives of a health promotion program must be realistic, governed as they are by the resources available to achieve them. For example, if behavior change is the objective, providing information alone, though less costly in personnel and materials than the pursuit of other goals, will not produce much of a change. In this case, a variety of educational approaches must be used, and more resources committed. How to set priorities in order to select and develop the most appropriate methods to achieve program objectives is the subject of the next chapter.

REFERENCES

1. Becker, M. H., ed. The health belief model and personal health behavior. *Health Education Monographs* 2(4) (1974).
2. Wallston, K., and Wallston, B. S., eds. Health locus of control. *Health Education Monographs* 6(1) (1978).
3. Bandura, A. *Social learning theory.* Englewood Cliffs, N.J.: Prentice-Hall, 1977.

2

Designing the Program

BUILDING OBJECTIVES BASED ON NEEDS

Identifying Needs and Priorities

A logical first step in the planning process is to identify needs and priorities. Some companies have launched their health promotion programs by specifying objectives, presumably with the intent of identifying health needs or problems on the basis of those objectives. But, as noted in Chapter 1, it is more prudent to proceed first with an analysis of employee health needs and then to specify objectives on the basis of the highest-priority needs. For the purposes of this discussion it is assumed that this is what your company intends to do.

Here are some of the questions an assessment of health needs should answer:

- What are the costs of medical care, disability benefits, and insurance premiums for the employees and their dependents?
- What are the use patterns for medical services by the employees and their dependents?
- What are the health problems of the employees and their dependents that contribute to the costs, disabilities, and use patterns identified above?

- What health promotion activities are currently being offered to employees? How appropriate are these in changing risk factors associated with the health problems identified above?
- What are the sociodemographic characteristics of the employee population that might relate to the health problems, risk factors, and behavior identified in the preceding queries?
- Do the employees see a need for a health promotion program?
- What do the employees think should be included in such a program?

As the proposed objectives become more ambitious or complex, so does the data collection required by the program. Two examples illustrate this:

1. *Proposed company objective:* Increase general awareness about healthy life-styles.

   Method of assessing need: Simple sampling of employees by questionnaires to determine their perceived health needs and their health interests.

2. *Proposed company objective:* Reduce the prevalence of cardiovascular disease.

   Method of assessing need: Compile complex data based on hospitalization, use of medical services, disability payments, and incidence of cardiovascular disease, as well as the prevalence of risk factors.

The objective of determining the cost benefits of a program would add still another layer of data requirements.

These examples indicate a direct relationship between the degree of complexity of a company's health objectives and the complexity and amount of data that will be required to pursue these objectives and to evaluate their achievement. A matrix illustrating the data needed to support certain program objectives is shown in Table 2.1.

Data That Could Be Collected

The data to be collected would ideally involve not only employees but also dependents and retirees.

Health Promotion Activities Currently Offered

- Identify the kind of health promotion activities currently being offered (e.g., hypertension control, exercise, smoking cessation).

- Identify how the activities are delivered (e.g., outside organization or in-house staff).
- Identify the types of educational approaches being used (e.g., behavior modification, small group discussions).
- Determine whether the current activities are meeting employee needs.
- Assess community resources (e.g., availability, accessibility, cost).

TABLE 2.1 Data requirements for health promotion objectives

TYPES OF DATA TO BE COLLECTED	PROGRAM OBJECTIVES						
	Increase in awareness	Increase in knowledge	Change in attitudes	Change in behavior	Reduction in cost	Reduction in risk	Reduction in morbidity and mortality
Health promotion activities currently offered	•	•	•	•	•	•	•
Effectiveness of current programs	•	•	•	•	•	•	•
Employees' perceived need for program	•	•	•	•	•	•	•
Employees' sociodemographic characteristics		•	•	•	•	•	•
Employees' current health habits			•	•	•	•	•
Prevalence of risk within employee population			•	•	•	•	•
Clinical indicators of risk or disease				•	•	•	•
Costs of medical services					•	•	•
Disability benefits and insurance premiums					•	•	•
Incidental and disability absenteeism rates					•	•	•

Effectiveness of Current Health Promotion Programs

- Determine the effectiveness of specific program components (e.g., hypertension control, smoking cessation, weight control) in relation to the objectives to be met. Use observation, questionnaires, clinical lab testing, etc.
- Determine the effectiveness of the overall program. What percentage of the population has been reached? Has there been significant risk reduction? Do changes in behavior accord with known standards of acceptability (e.g., 20–30 percent smoking cessation, 80–85 percent control of hypertension)?

Perceived Need for Program

- Determine the number of employees (and family members and retirees) interested in participating in the program components.
- Determine the preferences for taking part in such activities through community resources and/or in-house ones.
- Determine convenient time frames.

Prevalence of Risk

- Determine the prevalence of risk for specific diseases—for example, the risks involved in smoking and in elevated lipids. (See Addendum A, pages 54–56, for names and addresses of sources for risk data.)
- Group the data and target specific, high-risk groups for risk reduction. By thus limiting the initial program components to involve high-risk persons, the potential for payoff is greatest. Later the program can be offered to lower-risk groups.

Current Health Habits

- Determine dietary, exercise, smoking, drinking, and other habits. Use observation, self-report measures (questionnaires, telephone interviews), clinical lab tests, etc.

Sociodemographic Characteristics

- Determine such characteristics as age, sex, ethnic origin, occupation, employment status, education, residence (home and office). Use a health history or risk-appraisal form.

Clinical Indicators of Risk or Disease

- Determine blood pressure, height, weight, lipids, blood sugar, and other factors that might point to the risk or existence of chronic disease.

Cost of Medical Services

- Determine the number and cost of claims paid for hospital room and board, surgical procedures, physicians' office and hospital services, diagnostic procedures and X rays, outpatient emergency care, psychiatric care, prescription drugs, dental services, and vision care.
- Determine the utilization rate of hospital rooms per 1,000 persons per year.
- Average the medical expenses per year for employees, dependents, and retirees.

Disability Benefits and Insurance Premiums

- Determine the frequency and severity of disabilities.
- Determine the short- and long-term costs.

Incidental and Disability Absenteeism Rates and Costs

- Determine the number, frequency, duration, and costs of incidental and disability absences.

ORGANIZATIONAL LOCATION

The location for administering the health promotion program has a great bearing on how efficient and effective the program will be. The most frequent locations are the medical, the benefit, and the personnel departments. Some of the reasons for choosing any of these locations, the potential problems involved, and the possible solutions are presented in Tables 2.2, 2.3, and 2.4 (pages 27, 28, and 29).

A number of other locations are of course possible. Your choice of location hinges on the particular needs of your organization. For example, a company intending to market health promotion packages or products might decide on the marketing department as the most suitable location.

TABLE 2.2 The medical department as the administrative site of a health promotion program

PROS	CONS	POSSIBLE SOLUTIONS OF PROBLEMS
For those companies with medical departments, a health promotion program might be integrated into an existing structure.	Many medical personnel have far more extensive educational preparation and experience in the delivery of health care than in developing health promotion programs or evaluating them.	Obtain continuing education for in-house staff on how to develop, implement, and evaluate a health promotion program.
		Hire a professional suitably prepared to manage the program.
		Contract with outside groups to deliver various components of the program and/or make referrals to community resources.
	The location of the medical department within the overall company managerial hierarchy may seriously limit resource allocation for the health promotion program.	Obtain commitment of upper-level management for additional financial resources based on possible later cost savings, good national publicity, good recruitment aid, etc.
Medical personnel are available to conduct, consult, and/or provide immediate referral.	Health promotion activities often have to be accommodated to the schedules of medical personnel.	Modify job descriptions of company health professionals to provide a percentage of time devoted to the health promotion program.
Health promotion program efforts can be integrated with the ongoing occupational safety and health activities.	If the occupational safety and health division is separate from the department, the integration of a health promotion program becomes more difficult.	One objective of the health promotion program could be to better integrate the medical department and the occupational health and safety division.
The image of a company's medical department can be significantly enhanced by an employee health promotion program.	A negative image of a company's medical department may adversely affect participation.	It is a definite possibility that the image of the medical department will be enhanced as a result of the health promotion program. (An appropriate evaluation tool might be developed to measure change in employee attitudes toward the department.)
Regular clinic appointments can be linked with the health promotion interventions, thereby giving some employees time to participate who otherwise might not be able to do so.	Health promotion activities often have to be accommodated to the schedules of medical personnel.	

TABLE 2.3 The benefits department as the administrative site of a health promotion program

PROS	CONS	POSSIBLE SOLUTIONS OF PROBLEMS
Benefit planners are directly involved with the company's health care expenditures, and have access to existing health care data.	Sophisticated systems of collecting health care data often have not been developed by the company. Insurance carriers usually do not collect or store health and cost data in a manner that would facilitate its access by companies developing health promotion programs.	Large corporations that have data collection systems can help smaller companies develop data collection instruments and cost analysis formulas. The larger companies can initiate cooperative study programs with insurance carriers to determine the impact of health promotion programs on medical costs.
A cooperative relationship between the benefits and medical departments fosters the use of medical personnel for implementing a health promotion program.	The medical and benefits departments may lack a cooperative relationship, or even be in widely separate locations.	The medical department could begin to work with the benefits department on cost control measures.
Health care programs and policies are integrally related to benefit planning activities.	A health promotion program implemented by the benefits department may be viewed by employees as a way of reducing costs or increasing productivity, rather than as evidence that the company is interested in their health and well-being.	There is a need to involve employees in the process of planning, implementing, and evaluating a health promotion program. Such participation gives everyone a stake in the program and should reinforce commitment.
	Benefit planners may not have enough expertise to deal with health promotion effort. Outside sources, such as medical experts, would have to be used extensively for planning and implementation.	

TABLE 2.4 The personnel department as the administrative site of a health promotion program

PROS	CONS	POSSIBLE SOLUTIONS OF PROBLEMS
The personnel department is the initial contact for new employees and is often viewed as a unit concerned with the employee's welfare.	The personnel department may not be visible to employees after they enter the company. They might not associate health promotion with this department.	This department should be selected only if the staff of the personnel, medical, or benefits departments can share the responsibility for administration and implementation.
The personnel department may be the primary link for management with the unions, particularly in smaller companies.	The personnel department may be viewed as representing management, and therefore have a tenuous relationship with employees (and unions).	The personnel department can involve labor in identifying needs and implementing the program.
The personnel, medical, and benefits departments are all integral to maintaining the employee data that are necessary for the initiation and evaluation of health promotion programs.	The personnel, medical, and benefits departments may be organizationally independent of one another. The history of these departments in maintaining the confidentiality of employee data may affect programming efforts.	Delegate responsibility to the staff of only one department for health promotion. Use other staff as resources or consultants.
	Time and staffing constraints are a problem for implementing programs in this department. Health promotion activities may be viewed as ancillary to the primary work activities of a staff person. Personnel managers may require expertise in health promotion.	In most cases the health promotion program should not be *conducted* by personnel employees. However, they can serve as appropriate administrators.

IMPLEMENTATION

The implementation of a health promotion program involves decisions concerning organizational location, allocation of resources, and types of educational approach to be used. In addition, the practical issues of timing, publicity, and incentives, and even some ethical issues, require careful consideration. Although some of these decisions are made in the planning phase, others influence the types of approach that might be used for implementation. A consideration of some key implementation decisions—their pros and their cons—is presented below.

Methods of Implementation

Among the many approaches to implementation, the three most common are:

1. The pilot approach
2. The phased-in approach
3. Immediate implementation of the total program

Each of these has certain advantages and disadvantages, resource requirements, and expected outcomes. The pilot approach is often combined with either of the other two. The following description of these approaches should help you choose the one(s) most appropriate to your needs.

Pilot Approach

The purpose of the pilot, or demonstration, approach is to determine the feasibility of implementing a large-scale program. Are the appropriate data being collected? Is the proper method of delivery being used? In short, is the program, or are its components, clearly successful?

Whether the pilot involves a complete program or only program components (e.g., hypertension control, smoking cessation, weight control), it is directed toward a target group, is usually conducted in specific sites, and is evaluated to determine if the objectives have been met.

Advantages

- The components of a program can each be tested and, if necessary, revised before major programmatic and financial commitments are made.
- Specific educational methodologies (e.g., counseling, behavior modi-

fication, group discussions, lectures, mass media) can be tested for their appropriateness to program components.

- A cost estimate per employee and per program component can be made, as well as identification of resources needed to deliver each component.
- Close administrative control can be maintained by carefully choosing a site near the health promotion administrator.

Disadvantages

- The capacity to generalize the program's outcomes for a larger population is reduced if the characteristics of the pilot population and program components are too limited (as by sex, age, disease risk) or not representative (as with highly motivated volunteers who participate).
- Not all employees will be able to participate in the pilot.
- It may not be possible to obtain data on disability, on morbidity, and on medical costs applicable to the long-term impact of the program.
- If only selected program components, like smoking cessation and weight control, are piloted, the program may be criticized for being too categorical or not meeting all the needs of employees.

Phased-in Approach

Many companies choose to begin implementation on a limited basis, phasing in the program over a specified period. A pilot may or may not precede such an approach, or the early phases may serve as pilots for later phases.

There are a number of ways to phase in a program. Some examples:

- The first year the program might include only smoking cessation and hypertension control; the second year, weight control and exercise/physical fitness; and so on.
- The first year the program might begin at the company's national headquarters and then be phased in over several years at other geographical locations.
- At first the program might be limited to management employees, with phasing-in for nonmanagement employees later.
- The number of employees in the program might be limited for a given period, after which the number could be gradually increased.

Advantages

- Program personnel should be able to cope quite easily with the work load because they are not faced with meeting unrealistic expectations should all interested employees decide to enroll in the program.

- A gradual investment of financial resources can be made.
- There can be ongoing feedback to upper-level management about the effectiveness of program components over specific periods and adjustments can be made where necessary.
- Employees have the chance to enter the program when they are ready to make life-style changes, thus increasing the likelihood that such changes will occur.

Disadvantages

- Program components must be chosen and ranked by priority, and so must the locations and employee populations. Thus, neither all employees nor all program components can be provided for at once.
- If components are phased in, the company may lose an integrated approach to health promotion.

Immediate Implementation of Total Program

To implement a full-scale program at one time for all or a great many of its employees, a company must have the commitment of upper-level management and the necessary resources (money and personnel). This approach implies that a long-term (5- to 10-year) evaluation is being sought and that short-term evaluations will be used for program adjustments and not for determining the overall impact of the program on the company's health care expenditures.

Advantages

- All or a majority of the employees have the opportunity to take part in the program simultaneously.
- A long-term epidemiological study can be conducted with a large population data base.
- The resulting high national visibility can enhance public relations and recruitment.

Disadvantages

- The company may have to make a sizeable financial commitment that will probably include additional staff and plant facilities.
- Sophisticated evaluation measures, monitoring, and data analysis may be needed.

IDENTIFYING AND ALLOCATING RESOURCES

Personnel/Services

In staffing a health promotion program, you must consider the scope of the program, its organizational and geographical location, the time of day various program elements are to be offered, and the size and type of employee population to be served. Comprehensive programs require specialized health professionals and a staff approach different from that of programs that are sporadic or have more general objectives. During the planning phase, carefully assess the internal and external resources for their availability and cost. Resources for program implementation should be expended on the basis of two criteria: the budget allocation and the educational or programmatic specifications.

Internal Resources

There are specific advantages to using internal personnel and material resources: convenience, monitoring ease, flexibility when change is needed, ongoing educational reinforcement, availability, and continuity. Program components can be implemented on a regular basis by company staff as part of their job. The educational reinforcement can be completed by these professionals on a regular follow-up basis. If staff acquire additional skills to conduct program components, the company should receive a sound return on this educational investment. However, using internal resources may also mean redefining job descriptions, providing compensation for programs conducted outside of business hours, or hiring additional staff.

The actual reporting and supervisory relationships will vary according to the company and the department to which the health promotion activities are assigned. The usual procedure is for the health promotion director/specialist to report to a medical, benefits, or personnel director or vice president.*

Examples of Internal Personnel Resources

- Health professionals: doctors, nurses, health educators, behavioral scientists, exercise physiologists

*For a copy of a job description for the director or coordinator of a health promotion program, write to: National Center for Health Education, 211 Sutter Street, San Francisco, CA 94108.

- Employee assistance personnel: psychologists, psychiatrists, social workers
- Employees: managment/nonmanagement volunteer leadership, peer support groups, retirees
- Labor unions
- Trainers: training department
- Media personnel: public relations, graphics, television/video

External Resources

External personnel resources may have the flexibility of being available not only during but outside work hours. The costs will vary with the type of agency (voluntary, nonprofit agency vs. consulting, profit-making firm) that you decide to use. The scope of the activities and the length of program involvement will also influence costs. External resources are useful if staff resources are extremely limited and the budget decisions favor the use of a paid outside consultant or free resources.

External resources can, of course, be helpful in developing internal resources. For example, the American Cancer Society offers start-up smoking cessation programs and encourages the use of successful participants as volunteer leaders for ongoing company programs. Universities have similar programs for developing in-house staff capability—for example, health promotion training for a company's occupational physician(s) and nurse(s). See Addendum A (pages 54–56) for a representative list of both external and internal personnel resources and for sources of further information.

The primary disadvantage of using outside personnel and materials is the long-term costs associated with maintaining health behavior changes. Many community service agencies cannot supply the ongoing reinforcement necessary for changes in health habits. Further, since outside groups often are not oriented toward evaluation, they may lack familiarity with the methodologies of monitoring a health promotion program to ensure that the objectives are met.

Examples of External Personnel Resources

- Community health professionals: doctors, nurses, health educators, exercise physiologists, behavioral scientists, psychologists, social workers
- Health care organizations and institutions: hospitals, clinics, health departments, voluntary agencies
- Personnel from health promotion programs in other companies

- Consultants from universities, colleges, or the private sector
- Community agencies: American Heart Association, American Red Cross, Y's, among many others

In many communities there are a number of profit-making and nonprofit organizations and groups available to assist in developing and conducting a health promotion program. There are, for example, some 40 companies across the nation that are in business solely to provide corporations with such assistance. Some offer a wide range of services—from architects who design company gyms to consultants who conduct stress seminars *(1)*.

Often questions arise on how to evaluate the capability of outside resources. Addendum B offers a list of questions that may be helpful in assessing them (see pages 57–59).

Continuing Education for Personnel

In many cases internal personnel are an excellent resource for program implementation. But the complexities of a program, especially if it emphasizes behavior change and risk reduction, often necessitate a continuing education program for in-house staff. As part of the planning process, you should clearly assess the skill level of your staff and allow enough time for their continuing education before initiating the program.

If you use outside contractors to staff and implement the program, some kind of orientation program should be adopted for the in-house staff who will coordinate the effort. In this way they can be kept abreast of the latest methodologies in health promotion and better supervise the contractual arrangements. Even if outside groups are responsible for implementing part or all of the program, you may decide to initiate a maintenance component run by your own staff. Again, the in-house staff might have to increase their skills in this area through a continuing education program.

Facilities

Like staff resources, facilities can be provided internally and/or externally. An assessment of internal resources should be conducted, especially in regard to special equipment, space, accessibility, and convenience. If company facilities need to be supplemented with community resources, such as local schools, universities, and Y's, these external resources should be assessed to determine how they might be utilized.

Educational Materials

Today a wide range of educational materials is available for use in health promotion programs. Those chosen should closely match the program objectives. The reading level of the population to be served must be kept in mind when selecting reading materials. Preferably, they should be pretested with a target group.

The communications media are not a substitute for personal interaction. However, a number of instructional modalities (e.g., interactive computer systems, audio and video formats) permit a high degree of self-instruction. Both print and audiovisual materials should be used to trigger or reinforce a message. In and of themselves, educational materials cannot be expected to change a person's attitudes or behavior. But they may increase awareness or improve knowledge. The types of educational material available for health promotion programs are:

- Print materials: brochures, pamphlets, books, games, self-instruction kits
- Nonbroadcast media: local/national magazines, journals, newspapers, newsletters
- Broadcast media: films, videotapes, videodiscs, television (including cable), satellite
- Interactive media: computers used alone or in conjunction with other media (e.g., audio/slide, videotape)

Sources of educational materials are the federal government, insurance companies, professional societies, voluntary agencies, and private profit-making firms. Many of these materials are inexpensive, are free, or can be rented. Use of free materials can of course help keep program costs low. In the event that existing materials do not meet your company's needs, tailor-made materials are a possibility. In-house or community experts may be needed to ensure that high-quality materials are produced.

Educational Approaches

It is becoming increasingly evident that no single educational or behavioral intervention provides the final answer in assisting persons to change their life-style. Rather, the research literature suggests that a combination of educational and behavioral methodologies may be necessary and that the same combination or a different set of strategies may be in order to help maintain those changes. In this context, tailoring a health promotion program to meet the individual needs of employees takes on broad dimensions (see Table 2.5).

Large Group Meetings

The primary purpose of large group meetings is to convey information, heighten awareness, and provide opportunities for questions and answers.

- Suggested locations: conference rooms or auditoriums—with good acoustics and audiovisual capabilities—that are easily accessible to the chosen population; or rooms that are on employee routes to the cafeteria, elevators, stairways, post office, cashier, lockers, parking lot, etc.

Small Group and/or Person-to-Person Sessions

The primary purpose of small group or person-to-person sessions is to provide opportunities for peer support, role playing, simulation exercises, individual skill development, and behavioral therapy approaches.

Possible locations include any private office or clinic or conference room that is conducive to having discussions and promoting interaction between employees.

Time of Activities

A key issue in implementing the program is accessibility and convenience for the employees. Should the program be offered during and/or outside work hours? Each time period has its pros and cons.

Program Offered on Company Time

Pros

- It may be possible to involve employees who might not otherwise participate owing to family or other outside commitments.
- The timing demonstrates to employees the degree of interest the company has in their health and well-being.

Cons

- The timing may pose difficult logistical and financial problems where there are production shifts or a high number of participating hourly/nonmanagement personnel.

TABLE 2.5 The educational program

APPROACH*	DELIVERY CAPABILITY	MATERIALS	OUTCOME	
All employees	Media campaign	Audiovisuals: film, TV, videotape, slide/sound Visuals: exhibits, posters Print: bulletins, brochures, pamphlets, newsletters	Increase in awareness	
Large group	Program introduction Program recruitment	Audiovisuals: film, TV, videotape, slide/sound Visuals: slides, charts, bulletins, newsletters Print: brochures, pamphlets	Increase in awareness Increase in knowledge Change in attitudes	
Small group	Skill training Self-help Behavior modification	Print: self-instruction kits, workbook Audiovisuals: videotape, slide/sound, filmstrip, film Visuals: transparencies, slides, flip charts, chart pads, videodisc, audiodisc, games	Change in behavior Reduction in risk	Reduction in cost, morbidity, and mortality
	Learning seminars	Same as above	Increase in knowledge Change in attitudes	
Individual	Person to person Computer teaching Workbooks	Interactive: computers, teaching machines Print: self-instruction kits, programmed instruction, pamphlets Same as above	Increase in knowledge Change in attitudes Change in behavior Reduction in risk	Reduction in cost, morbidity, and mortality

*Imaginative programs are created through use of more than one approach, combined with well-designed and tested materials.

- Work schedules of management personnel frequently conflict with the program.
- In all cases, employees may have to negotiate with an immediate supervisor for time away from the job.

Program Offered Outside Work Hours

Pros

- Busy work schedules are usually not interrupted and program costs can be kept to a minimum.
- The program tends to attract only those who are really committed to the effort.

Cons

- Employees may be unwilling to give up their limited free time (e.g., a lunch break) to attend. This is especially true when the program is several weeks long, or takes place in a major city where commuting may be a problem, or is conducted at a time of year when holidays or vacations intervene.

Incentives

A number of companies are offering incentives to encourage employees to participate in health promotion programs. For example, some companies may pay overweight employees a certain sum for each pound they shed. Other companies offer monetary rewards for the number of miles a person jogs, bikes, or swims. But this method should be used with discretion. As Jim Fixx notes:

> Some time ago Jess A. Bell, who heads the Bonne Bell cosmetics firm, decided he wanted to encourage his employees to exercise more. A practical man, Bell devised what seemed to him a straightforward scheme. He offered a dollar for every mile run, fifty cents for each mile walked, and twenty-five cents for each mile covered on a bicycle. The program succeeded so well that, paradoxically, Bell was forced to rate it an utter failure. "We ran out of money," he says with dismay. "There were people running 200 and 250 miles a month. They obviously weren't out selling—they were out running. So we had to give that up." *(2)*

Extra 30-minute breaks at lunch and extra vacation days are other incentives being tested by companies to obtain the active participation of employees in the program and to bring about health behavior changes. For the most part, incentives can be divided into three categories: released time, compensation, and awards (e.g., T-shirts, plaques, publicity).

Though incentives are usually used as rewards at the end of successful completion of a program component (e.g., exercise, smoking cessation, weight loss), they may also be used during a program either for positive reinforcement or to maintain behavior changes once the program itself is over.

Several ethical issues are involved in the use of incentives. Some persons may be healthy to begin with; others may have a genetic condition that can be modified only in a limited way or perhaps not at all. What incentives can be offered to such persons?

In addition, the measurement of and rewards for behavior change are difficult issues for persons making changes on their own, in contrast to those who are enrolled in a program sponsored and monitored by the company. Another problem: After behavior change has been achieved, does the company continue to provide incentives so as to reward health maintenance? What happens if the person relapses to old habits?

Publicity

The use of publicity is of course necessary in implementing the program, but it is not nearly as critical as a proper needs assessment conducted in the planning stage. Poor publicity, though usually cited as the reason for program failure, is usually not the reason. More likely causes of failure are a poor assessment of employee needs, attitudinal or belief patterns, improper timing, inaccessibility, lack of reinforcing factors such as supervisor acceptance, or disinterest in the subject chosen.

Forms of publicity available are:

- House newsletters
- Desk drops
- Payroll stuffers
- Internal video systems
- Special invitations by mail or phone
- Posters and signs
- Outside sources: newspapers, magazines, radio, local cable television, posters, etc.

Internal resources are:

- Computers

- Public relations
- Media or graphics department
- Employee/retiree associations and clubs
- Union leadership

Funding

The funding of health promotion programs varies with the organization. There are three basic methods. Programs and materials can be provided (1) at cost, possibly on a tuition/reimbursement basis, (2) on a cost-sharing basis, or (3) at no cost.

In the first method the basis for costs can vary greatly. For instance, costs can be limited to incremental ones, like instructor fees and course materials, with no charge for developmental expenses or facilities if the program is conducted off company premises. Or costs can be fixed and worked out on a tuition-refund basis. In this case, the employee must pay the money in advance, which could hinder participation. This approach also requires that program rules be established and fees set. Further, it must be decided what percentage of attendance is required for an employee to have successfully completed the course and be entitled to reimbursement. Any exceptions (e.g., illness, business travel, company transfers) should be clearly specified in advance. The money could also be refunded on a time-sequence basis as a reward for maintaining behavior change.

If the cost-sharing method is used instead, fees per course should be established to ensure proper cost control. Such fees will depend on the company's ability to pay, the degree of co-payment established, and the range of charges for similar programs in the community. Taxation issues should be analyzed. There are many angles to the cost-sharing method. Each company has to work out its own formula.

Whatever your funding method, carefully consider such issues as equality, consistency, the employees' ability to pay, and the degree to which the employees will be motivated to a healthier life-style by incurring all or part of the costs.

CONCLUSIONS

Health promotion does not, unfortunately, come in a neat package. In implementing a health promotion program, always keep in mind what can reasonably be delivered and emphasize the individual's responsibilities in the activities. Whenever possible, involve the employees in orchestrating all phases of the program—planning, implementation, and evaluation. With this kind of participation, both interest and commitment can be intensified.

REFERENCES

1. Carroll, Jane. Exercising the choice of good health. *San Francisco Examiner*, November 24, 1980, p. C-1.
2. Fixx, James F. *Jim Fixx's second book of running.* New York: Random House, 1980, p. 50.

3

Evaluation

NEED FOR EVALUATION

The two questions that arise most frequently concerning the evaluation of health promotion programs are:

1. What evidence is there that these programs will improve employee health?
2. What evidence is there that these programs are cost-effective?

In attempting to address these questions, the first inclination is to study existing company programs to determine if there are any models or program elements that can be adapted to other settings. The difficulty with this approach is that only a few of those programs have had an evaluation component. And even where an evaluation design is available, there are as yet usually no definitive data on risk reduction, on changes in morbidity and mortality, or on cost-effectiveness.

DEFINITION OF EVALUATION

Evaluation can be defined as "the comparison of an object of interest against a standard of acceptability." The object of interest is the program or program objective; the standard of acceptability is the value to which the program is

This chapter is based on L. W. Green et al., *Health education planning: A diagnostic approach* (Palo Alto, Calif.: Mayfield Publishing Co., 1980), with permission of the publisher.

compared; and the method of comparison is the way the program objective is analyzed in relation to the acceptable standard.

Evaluation of health promotion programs can therefore be viewed simply as comparing an object of interest (increased knowledge, behavior change, or risk reduction) against an acceptable standard (say, 25 percent smoking cessation, 80 percent compliance with regimens for high blood pressure control).

PURPOSES OF EVALUATION

From a business point of view, one of the primary purposes of evaluation of a health promotion program is to assure that monies being allocated for the program are used in the most efficient and effective manner. From the standpoint of accountability, the main purpose is to determine whether the program objectives are being met. Evaluation also adds to the body of scientific and practical knowledge that can be diffused to employees and employers alike so that they can make informed decisions about life-style changes and health care.

Although most of a program's decision-makers agree that evaluation is important, they often exclude specific evaluation activities because they feel either that evaluation is too complex or that the resources necessary to conduct acceptable evaluation are not available. Evaluation need not be too complex or too expensive. A range of evaluation methods and designs is available to industry, the choice depending on specific program objectives (e.g., increased awareness, behavior change, risk reduction), the type of comparisons desired (e.g., changes over time, changes between groups, changes compared to mandated standards), and the resources available to conduct the evaluation (simple recordkeeping, sophisticated statistical techniques, epidemiological assessments, etc.). More will be said about the types of comparison possible and how acceptable standards can be decided on.

LEVELS OF EVALUATION

Traditionally, health program evaluation has focused on two levels of measurement—process and outcome. *Process evaluation* (or formative evaluation) focuses on the manner in which program activities were conducted and on methods of assessing the quality and appropriateness of the professional practice. Examples of process evaluation measures are: the number of life-style brochures distributed, the quality of information on physical fitness programs, and the type of instructors or teaching methods used.

Outcome evaluation measures the consequences that are attributable to program activities. Examples of outcome evaluation measures are: changes in knowledge, in attitudes, and in behavior; risk-factor profiles; and decrease in morbidity and in premature mortality.

A program can have a short-term impact on the knowledge, attitudes, and behavior of the participants, as shown by outcomes such as:

- Number of courses requested
- Number of sessions attended
- Improved knowledge of cardiovascular risk factors
- Increased belief in the preventability of chronic disease
- Increased levels of regular aerobic exercise
- Less cigarette smoking
- Less consumption of high-cholesterol foods
- Greater adherence to regimens to control high blood pressure

Additional outcome measures may be associated with a program in the long term but may be difficult to attribute directly to the program itself. Examples of long-term outcome measures are:

- More productivity
- Less absenteeism
- Fewer health care claims
- Less cardiovascular morbidity and mortality

At this stage in the development of the programs, evaluation efforts that focus on the short-term outcomes of a program appear to be the most useful and most needed. These are the outcomes of activities that encourage preventive behavior (e.g., smoking cessation, increased physical activity, and stress management).

Documentation of the process by which health promotion programs are conducted is necessary but not sufficient to answer the questions posed at the beginning of this chapter. Emphasis on the long-term effects of the programs is extremely difficult, given the practical constraints of most of the programs operational in industry. Validation of the hypothesized long-term benefits requires many years of sophisticated health promotion research involving randomized evaluation designs.

Regardless of the level of evaluation (process or outcome) or of the basis of measurement (short-term or long-term), the extent of measurement depends on the specific evaluation design and the standards of acceptability proposed for the health promotion program.

STANDARDS OF ACCEPTABILITY

In evaluating a program, an object of interest (usually an outcome measure based on a program objective) is compared to a standard of acceptability. The method of determining whether the object of interest has met a predetermined standard depends on the specific evaluation design and standard of acceptability selected.

For instance, if the object of interest is increased participation in a physical fitness program, a suitable standard of acceptability could be a 10 percent increase after six months in the number of employees participating in that program compared to baseline levels. If this has been shown to be a reasonable rate of increase in other health promotion programs in industry, it could then be considered to be a *normative* standard of acceptability.

Some other sources of acceptability are:

1. *Historical standards.* Current program outcomes are compared to results from a previous period. For example, say that last year's smoking cessation program yielded a 10 percent reduction in the number of employees who smoke. A historical standard of acceptability can be obtained by comparing the results of this year's program with last year's 10 percent reduction.
2. *Theoretical standards.* Program outcomes are compared to the level expected if everything goes as planned. The theoretical standard is often based on previous research in which the interventions have been tested in controlled laboratory or clinical situations. For example, a demonstration physical fitness program, conducted by university researchers using the latest in educational strategies, yielded a 40 percent regular participation rate among students. This could serve as the theoretical standard of acceptability, with the differences between students and employees taken into account.
3. *Absolute standards.* Program outcomes are compared to the highest possible level attainable. Whereas theoretical standards are based on the premise that everything will go as planned, absolute standards are often unrealistic and may never be possible to attain. For instance, a smoking cessation rate of 100 percent among employees, an example of an absolute standard, is not realistic.
4. *Negotiated standards.* Program outcomes emerge from the compromise and negotiation of other possible standards. A negotiated standard is frequently an average of the preceding standards of acceptability. For example, if other smoking cessation programs yield a 30 percent cessation rate (normative), historical standards are approximately 20 per-

cent, the theoretical cessation rate for your population is 40 percent, and the absolute standard is complete (100 percent) cessation, then a negotiated standard could be 35 percent—that is, a weighted average of the other standards that gives greater weight to historical and normative standards than to the absolute standard.

Once the standard of acceptability has been determined, the specific evaluation design needs to be selected. As was the case with standards of acceptability, there are several evaluation designs from which to choose. Your choice depends on your company's resources and program objectives.

DESIGNS FOR EVALUATION

Current health promotion evaluation is sometimes criticized for a lack of rigor in experimental design, absence of control populations, and evaluation measures that fail to discern changes in health behavior beyond changes in knowledge and attitude. Further, the short time frame within which most programs have been assessed has stymied the search for definitive answers.

It is possible, however, to attain a level of rigor in an evaluation that accommodates organizational and economic limitations while sacrificing very little validity. Your decision to use a specific design for evaluation must take into account not only the limitations of any design but also the practical, ethical, and financial constraints that influence the way you conduct your evaluation. For example, you may find it very difficult to achieve the ideal of a randomized population group because of informed-consent issues and the necessarily voluntary nature of the program. Similarly, it may not be possible for you to identify a group of employees to serve as a control group. Despite such limitations, there are a variety of evaluation designs that allow for assessment of the effectiveness of health promotion programs at different levels. Depending on your program's objectives, the available resources, and the need for more or less conclusive or compelling results, you can select a suitable evaluation design from the following examples.

Design A: The Historical, Recordkeeping Approach

In this approach, the evaluator simply sets up a recordkeeping procedure to accumulate appropriate data, and periodically charts the data to show the change that is occurring. How frequently the data are charted depends on how often the events that are being tabulated occur. This very simplistic approach yields graphs and charts that effectively demonstrate how the program is doing.

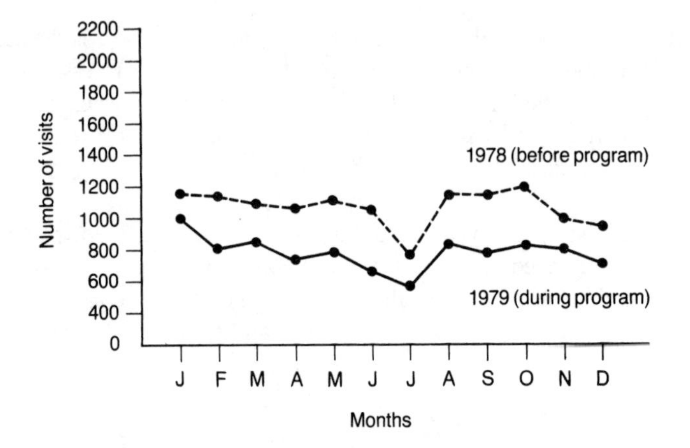

FIGURE 3.1 Corporate health services/total dispensary sickness visits [Kimberly-Clark Corporation]

Design A is illustrated by Figure 3.1, which the Kimberly-Clark Corporation constructed as an early evaluative report from routine records on participation in its health promotion program *(1)*.

Design B: The Inventory Approach

In this approach the evaluator must make a special effort to collect data periodically, at specific intervals, rather than accumulate them on an ongoing basis over the entire test period. So he or she must set target dates for assessments of the program, identify the expected target levels, and perform sample surveys or observations. For some programs the critical measurement points are standardized (e.g., smoking cessation at 1½, 3, 6, and 12 months).

Following are examples of programs (discussed more fully in Chapters 25, 26, and 28) that were evaluated through the inventory approach.

1. A high blood pressure control program, developed by Cornell University Medical College for the United Storeworkers Union in New York City, detected and treated hypertensive employees at Gimbels and Bloomingdale's. Since 1973, screenings have revealed some 3,000 employees of the stores to be hypertensive. Of these, about 2,200 receive care at one of 19 union-provided treatment centers. With a health team approach, a nurse, supervised by a physician, provides care according to a systematic protocol. There are no direct costs to patients for visits, drugs, or laboratory tests. Patients have adhered well to

treatment, with attrition amounting to less than 10 percent annually. Eighty percent of active patients have achieved and maintained satisfactory blood pressure control. Preliminary data collected on a systematic basis suggest that absenteeism and hospitalization have declined among the treated patients. (See Chapter 25.)

2. An oil production company in Texas requested that a local hospital's Weight Control Clinic conduct two series of weight control classes for the company's employees. The company paid for the program and allowed time off from work to attend eight one-hour classes. A dietitian familiar with behavior modification techniques and nutrition information conducted the classes for two groups of employees. Baseline diet and behavioral questionnaires were completed on each employee, along with such data as weight, height, and triceps skinfold measurement. At the end of eight weeks, data were collected again for historical comparison *(2)*.

3. The Campbell Soup Company (Camden, N.J.) developed a working agreement with a local university to conduct a series of smoking cessation classes. The program model was one in which a service was provided, data were collected periodically rather than continuously, and staff were trained at the firm so that in-house programs could be undertaken by the company itself but evaluated by the university. A six-month follow-up survey found that 25 percent of participants were still abstinent *(3)*.

Design C: The Comparative Approach

This design is one in which the standard of comparison is evaluation efforts that have been completed in other settings. An evaluator usually can identify data on similar programs in other places and can then borrow or buy the standardized format for collecting data or keeping records. He or she then uses these forms and makes periodic comparisons on the same basis as Design A or Design B. AT&T recently developed a standardized evaluation protocol for its smoking cessation programs. The protocol includes a smoking history and follow-ups at intervals of 3, 6, and 12 months. It is designed for use throughout the Bell System, so comparisons can be made of the success rates of programs run by Bell System personnel and/or outside resources.

Data from a particular program can also be compared with national data. For example, the National Health Survey provides standardized questionnaires from which data can be used for such comparisons. Standardized formats for data collection are recommended because comparability with another test or control population is often an essential component of cumulative evaluation. National surveys provide national norms and show what should be expected in health-related behaviors for populations defined by age, sex, and educational level. Thus any evaluation of a worksite health promotion program could be compared with national or regional data or previous studies in the literature. As Danaher notes in Chapter 28, the odds are 7:3 that any participant in a smoking cessation program will relapse within three months.

Design D: The Controlled Comparison, or Quasi-experimental, Approach

In this approach, the evaluator identifies a community or a population that is similar to the company's target population but is not participating in a health promotion program. Design A or B is applied to both the target population and the comparison population, which are then periodically compared at about the same time.

The Stanford University three-community study is the most notable example of this approach. Three communities were surveyed: one in which an intensive, all-out health education effort had been made; one in which a mass media effort had been made; and one in which neither effort had been made but which had comparable resources and facilities. The educational strategies used in the three communities and for their various subpopulations were compared using short-term outcome data obtained from surveys *(4)*.

Another example of Design D is a Ford Motor Company project in which three distinct intervention programs were established for hypertension. These included two methods for detection and referral and one for on-site treatment at separate automotive plants. No intervention is planned for a fourth plant ("control" site) so that results can be compared with those of the experimental locations. (See Chapter 9.)

Design E: The Controlled Experimental Approach

This approach is comparable to the clinical trial in medical research. The evaluator establishes a formal procedure for randomly selecting the persons within the study population who will participate in the experimental health promotion program and those who will not. This approach requires a situation in which it is possible to deny some persons the experimental health promotion intervention. The evaluator collects the data in both the experimental and the control groups to track their progress.

One example of the controlled experimental approach is a cooperative study that was conducted by AT&T and the National Cancer Institute (NCI) to evaluate the NCI's informational program "Progress Against Breast Cancer." Women employees at several AT&T sites were invited to take part in the study. Participants were randomly assigned to four experimental groups receiving the educational program. Changes in knowledge, attitudes, and behavior and the use of different teaching methods were measured in the four groups. Control groups at two other AT&T locations were used to monitor major historical events that might affect the study results. Control group participants were given the educational program after the study measurements were taken.

Design F: The Full-blown Evaluative Research Project

This approach is not feasible for most employee health promotion programs *(5)*. The strategies from Design E are applied but, in addition, multiple groups are randomized to systematically varied combinations of program elements, and multiple measurements are obtained. Each group receives a different mix of health promotion interventions (e.g., Group A, smoking cessation alone; Group B, smoking cessation and physical fitness; Group C, physical fitness alone; Group D, no program).

For example, a Johns Hopkins health education study on hypertension randomized 400 patients to eight groups that represented variable exposure to eight combinations of three different educational interventions. Subgroups were then compared by age and sex to determine the effects of the various combinations of education on different types of patients. Significant reductions in blood pressure were obtained with most combinations of two or more educational interventions *(6)*.

Another example is provided by SRI International's project in collaboration with the Lockheed Missiles and Space Company and Kaiser Permanente. This was a multifactorial study of the effect of educational interventions, barriers to health care, and behavior modification of compliance to antihypertensive treatment. Hypertensives were screened at their worksites, with the use of multiple blood pressure assessments, and were then randomized to an educational intervention, a barrier reduction (e.g., convenient appointments, telephone reminder calls), and a behavior modification intervention. The results supported findings that educational materials alone do not increase compliance, whereas barrier reductions were significantly effective. In addition, behavior modification proved to be effective for those hypertensives who had the most difficulty in adhering to their medical treatment *(7)*.

CHOOSING AN EVALUATION DESIGN

The more rigorous designs are time- and labor-intensive and may require informed consent. But the main problem with them is that they usually have to be applied under highly controlled conditions in which the behavioral circumstances are unusual or unnatural. Thus they tend to remove people from the very context in which their behaviors ordinarily occur. The research findings from a highly controlled laboratory, classroom, or clinical facility may not be easily applicable to the variety of community- and industry-based programs. In short, a study may gain validity through the use of more rigorous procedures but may sacrifice feasibility and generalizability.

But somewhere between the simplicity of evaluation via Design A, which leaves you with suggestive but inconclusive findings, and the complexity of evaluative research via Design F, which is very difficult to conduct, there is an optimal level at which evaluation is both sufficiently rigorous and clearly practicable.

To determine which evaluation design to use, first select the program outcomes you want to measure and their corresponding standards of acceptability. Then choose an evaluation design that (1) provides the data necessary to determine whether or not your program is reaching its objectives and (2) is consistent with the resources available for evaluation.

In addition, derive some confidence from the epidemiological data base that links specific health habits (smoking, nonadherence to regimens for hypertension control) with diseases and illnesses (e.g., lung cancer, heart disease, stroke). For with these kinds of established linkages and others that will undoubtedly be substantiated in the future, you should be able to count on reasonable evidence of cost benefits and cost-effectiveness based on your short-term measurement of changes in behavior.

CONCLUSIONS

Preliminary evidence from some health promotion programs indicates that changes are occurring in health knowledge, attitudes, and behaviors, but it is too early to have definitive data on downward trends in cardiovascular disease, cancer, and stroke, in morbidity and premature mortality, and in medical costs. Additional evidence should be forthcoming the longer these programs are in place and as more of them include measures to evaluate their impact. In the meantime, many companies have now embarked on health promotion programs, viewing them variously as good for employer-employee relations, employee benefits, short-term payoffs to the company, and long-term financial investments in the health of employees.

REFERENCES

1. Dedmon, R., et al. *Kimberly-Clark's health management program interim report, March 1, 1980.* Neenah, Wisc.: Kimberly-Clark Corp., 1980, Fig. 22.
2. Foreyt, J. P.; Scott, L. W.; and Gotto, A. M. Weight control and nutrition education programs in occupational settings. *Public Health Reports* 95:127–136 (1980).
3. Wear, Roland, F., Jr., M.D., Corporate Medical Director, Campbell Soup Company, Camden, N.J. Personal communication, January 1979.
4. Farquhar, J. W., et al. Community education for cardiovascular health. *Lancet* 1:1192–1195 (1977).
5. Green, L. W. Research designs applicable to the practice setting: From rigor to reality and back. In *New dimensions in patient compliance,* edited by S. Cohen. Lexington, Mass.: D. C. Heath and Co., Lexington Books, 1979.
6. Levine, D. M., et al. Health education for hypertensive patients. *Journal of the American Medical Association* 241:1700–1703 (1979).
7. Chadwick, J. H., et al. Blood pressure education in the industrial setting: Notes for a progress report to the National High Blood Pressure Education Research Program. Menlo Park, Calif.: Stanford Research Institute, 1976.

ADDENDUM A

Personnel Resource Inventory

For current information on health promotion and available resources across the country, the following organizations should be contacted:

National Health Information Clearinghouse*
P.O. Box 1133
Washington, DC 22013

National Center for Health Education
211 Sutter Street
San Francisco, CA 94108

(Data Base)
Center for Health Promotion and Education
Centers for Disease Control
1600 Clifton Road
Atlanta, GA 30333

Following is a list of health promotion program offerings with the corresponding external and internal resources that might be tapped. This list is merely representative.

*The clearinghouse is sponsored by the Office of Health Information and Health Promotion, Physical Fitness and Sports Medicine.

PROGRAM	EXTERNAL RESOURCES	INTERNAL RESOURCES
Cardiac pulmonary resuscitation (CPR)	American Red Cross, colleges, hospitals, police and fire companies, rescue squads, Scouts	Each of these programs could utilize one or more of the following: ◆ Doctors ◆ Nurses ◆ Health educators ◆ Psychologists/counselors ◆ Nutritionists ◆ Exercise physiologist(s) ◆ Safety personnel ◆ Employees ◆ Social workers
First aid (including obstructed airway technique)		
Diabetes control	American Diabetes Association, National Institute of Arthritis, Metabolism, and Digestive Diseases	
Employee assistance program (mental health related problems, drug/alcohol abuse control)	Alcohol, Drug Abuse and Mental Health Administration, community mental health centers, colleges, hospitals, physicians, private organizations, psychologists	
Exercise/physical fitness	American Association of Fitness Directors in Business and Industry, colleges, President's Council on Physical Fitness and Sports, schools, Y's	
Healthy back	Colleges, hospitals, physicians, Y's	
Hypertension control	American Heart Association, local health departments, National Heart, Lung and Blood Institute	
Safety/accident prevention	Carriers, insurance companies, local safety councils, police, rescue squad, workers' compensation	
Smoking cessation	American Cancer Society, American Lung Association, colleges, hospitals, National Cancer Institute, National Institutes of Health, private organizations and practitioners	

PROGRAM	EXTERNAL RESOURCES	INTERNAL RESOURCES
Stress management	Colleges, health practitioners, National Heart, Lung and Blood Institute, National Institute of Mental Health, private organizations	Each of these programs could utilize one or more of the following: ◆ Doctors ◆ Nurses ◆ Health educators ◆ Psychologists/counselors ◆ Nutritionists ◆ Exercise physiologist(s) ◆ Safety personnel ◆ Employees ◆ Social workers
Weight management/ nutrition	Colleges, hospitals, National Heart, Lung and Blood Institute, practioners, private organizations, Y's	

ADDENDUM B

Questions for Assessing Community Resources

Although most companies that have made a commitment to employee health promotion would probably prefer to use their internal resources to conduct these programs, many cannot, and must look for services from community-based profit-making and nonprofit organizations. To make maximum use of these resources, a company must know clearly what kind of assistance it is looking for, the frequency of use, and the associated costs.

NEEDS ASSESSMENT

Health promotion programs are designed to change behavior patterns known to increase the risk of disease. A company must therefore know what its employees' health problems are. When a company is working with community resources on a needs assessment, the following questions are relevant:

- Is the community resource offering to conduct the needs assessment?
- How will it do the assessment? Samplings, interviews, questionnaires?
- What evidence of prior experience does it have with such an activity?
- Can your public relations, marketing, or other research section conduct a needs assessment for equal or less cost?
- What are the confidentiality issues in using an outside resource?

PROGRAM OBJECTIVES

Probably the most common error companies make in using outside resources is not knowing exactly what outcomes they want their program to achieve. This makes them vulnerable to having their objectives set by a community resource that is unfamiliar with the company; indeed, the resource may have a set of objectives of its own that is congruent with its marketing approach.

There are short-term and long-term, simple and complex, objectives. At this point in the history of health promotion, changes in awareness and knowledge and improved morale are short-term and relatively simple objectives. Reduced health care costs and increased productivity are long-term and relatively complex objectives. Using simple informational techniques like films and pamphlets to achieve complex objectives related to behavior, costs, or increased productivity is inadequate. Unless your company knows exactly what outcomes it is looking for, using a community resource can be an unnecessarily expensive or unproductive endeavor. Rather than implement objectives offered by a community resource, it would be better for your company to set the program objectives itself and then seek the appropriate community resource to implement the program to those ends.

IMPLEMENTATION

Implementing a health promotion program requires decisions on two key questions. (1) Is the program's purpose to achieve basic awareness, behavior change, and/or risk reduction? (2) Is evaluation desired to determine whether the program was effective? For purposes of this discussion it is assumed that behavior change is the desired outcome and that the evaluation will concern itself with the program's effectiveness in producing behavior change and subsequent risk reduction. This means that there must be interventions designed to change established habit patterns plus reinforcement.

Here are implementation questions that can be asked of the community resources:

- Can they offer programs on a regular basis to you for six months to three years? If not, will they train your professionals, and at what cost?
- What is the level of experience and training of their personnel?
- Are they available before and after work and during lunch hours?
- What volume of participation can they handle at any one time?
- How is their cost determined—by the hour, by the day, per person? Is preparation time also charged for?
- How will the confidentiality issue be handled?

- Can you observe their services elsewhere in order to preview their skills much the same way you would a film?
- Does their contract include administrative overhead? If so, how much?
- Do they supply the materials? If so, at what cost?
- How reliable are they? (Look to other companies for past experience with them.)
- How much control over decisions do you have, want, and need?

EVALUATION

Evaluation is the most important part of a health promotion program other than a clear set of objectives. If a community resource is being used to design all or part of an evaluation strategy, its past evaluation reports should be checked carefully. Some of the questions that need to be asked are:

- What do they believe the evaluation should consist of? Process or outcomes? How do they measure costs?
- What have their past results shown regarding success rates, and what were the measurement periods?
- How did they evaluate? What methods did they use—subjective self-report measures like telephone interviews and questionnaires, or objective measures like clinical lab tests or observation of skills?
- How much does the evaluation cost? Is it being added on to their original cost?
- Who performs the data analysis? Who writes the final report?
- What importance do they attach to evaluation as part of their service? (If it's not important to them, you might want to use another resource.)
- Will they allow you to use your company's evaluation techniques—or do they want only their own used?

CONCLUSIONS

Using community resources for any or all the phases of your program may at first appear to be a simple way to facilitate matters. But this procedure is complicated, requiring careful decisions, clear objectives, and administrative coordination and monitoring. Nevertheless, the use of outside resources is often very worthwhile. The above checklists should help you use them effectively.

The health promotion programs of 17 large companies are described individually in Chapters 4 through 20. The programs vary considerably in their makeup, some being geared for a few thousand participants, some for many thousands. The descriptions therefore offer a basis of comparison that should be helpful to any company, large or small, that is setting up or revamping a health promotion program. The survey of each company program includes: sponsorship, objectives, target population, program structure, program processes, costs, evaluation, and special features.

The descriptions were compiled under contract to the National Center for Health Services Research, Department of Health and Human Services.

II

EXAMPLES OF COMPANY PROGRAMS

4

American Hospital Association

Sponsorship

The American Hospital Association offers the Well Aware About Health risk-appraisal screening and educational program to all AHA employees at their Chicago headquarters. The Well Aware About Health program was designed and developed at the University of Arizona.

Objectives

The primary goal of the program is to identify employees' health risk factors and to warn them of potential health hazards before symptoms of disease or illness occur. The program focuses on helping employees to change unhealthy behaviors through health information seminars and educational programs.

Target Population

The program is offered on a voluntary basis to all Chicago-based employees and their adult family members. Participating in the 1981 program were 360 employees (48 percent of the total work force) and 26 of their family members.

Used with the permission of the American Hospital Association.

Program Structure

Each Well Aware participant completes a health questionnaire that includes a personal assessment of his or her medical history, life-style, knowledge of health facts, and attitudes about health care practices. The employee then goes through a screening program that notes height, weight, and blood pressure and includes several blood tests and a urinalysis. A simple step test is conducted to determine the participant's relative level of fitness. Participants receive a personal health profile based on questionnaire responses and exam results that explains significant findings, pinpoints applicable risk factors, and offers recommendations for improving health status.

Health information seminars and risk reduction sessions are available to employees who want to change specific life-style behaviors. A series of health information seminars is conducted by local hospital health professionals to provide employees with information on how to stay healthy. Educational classes are conducted on such topics as fitness, stress management, smoking cessation, weight control, and nutrition.

Program Processes

Orientation sessions are held for employees who are interested in participating in the program. One month after the screening program is completed, each employee receives an individual health-risk profile. Employees identified as moderate or high-risk are encouraged to schedule individual counseling appointments with local hospital nurses who have been trained by Well Aware staff to provide educational counseling and recommend life-style changes.

Cost

AHA offers the Well Aware program as a free health benefit for all employees and underwrites all screening program costs. Employee family members are asked to pay a nominal screening fee.

Employees who participate in the educational classes are charged a fee to ensure their commitment to the program. As an additional incentive, those who attend a specified number of sessions of the fitness or smoking-cessation classes are refunded a portion of their class registration fee.

Evaluation

AHA receives an evaluation report prepared by Well Aware About Health that summarizes the *aggregate* screening results of all program participants;

it does not retain any data about individual employee results.

Employees who participate in Well Aware About Health are surveyed to determine their opinions of the program. Past responses indicate a high degree of satisfaction with the program content and structure.

Special Features

- Orientation sessions are held to familiarize prospective participants with program objectives and procedures.
- AHA has developed an incentive program to encourage employees from all levels within the organization to participate. To stimulate involvement in the program, interdepartmental competition is encouraged and prizes are awarded to departments with the highest percentage of participation.
- Employee representatives are selected from each department to help promote the program throughout the organization.
- AHA contracts with local hospitals to process all laboratory data and to provide staff to perform screening procedures and conduct educational sessions.

5

American Telephone and Telegraph Company and the Bell System

Sponsorship

Health promotion in the Bell System includes a variety of programs to increase awareness, knowledge, and the adoption of healthy life-styles. Program areas include health education, employee assistance, periodic exams, and, in a few companies, physical fitness. These program efforts are administered by the individual company medical departments and coordinated on a national level through the AT&T Corporate Medical Division. Many of the health promotion efforts are standardized throughout the System, while others are designed to meet the special needs of individual companies.

Objective

The purpose of the health promotion effort is to improve the health of employees by encouraging them to voluntarily adopt healthy life-styles.

Target Population

Health education and employee assistance programs are offered by the companies to all employees, and in some cases to their dependents. The physical fitness units are used for upper-level management and for those employees with disease conditions that could be positively affected by a planned exercise program.

Program Structure

Bell System health promotion programs are managed through the company medical departments. A coordinator in each company is responsible for health education program development in addition to regular nursing functions. Counselors in the employee assistance programs (EAPs) usually are on the Medical Department staff or part of contracted services from an outside firm. Exercise physiologists work closely with the medical personnel for the fitness units.

One or more of the following components are included in Bell System health promotion programs, although each company offers its own choice of components. Life-style programs focus on high blood pressure control, smoking cessation, weight control and nutrition, cancer education, physical fitness, and cardiopulmonary resuscitation. Employee assistance programs provide counseling and referral for substance abuse (alcohol, drug) and for emotional/behavioral problems. The periodic exam and health risk appraisal program provide the opportunity to intervene on health conditions and risks factors.

Health education programs utilize both internal medical staff and local community organizations. They are usually conducted outside of company hours and are available to all employees and, frequently, their spouses.

Program Processes

There are three methods whereby employees receive educational interventions to help them improve their health: (1) Employees using the various company health clinics complete a medical hstory or health risk form. They are counseled or referred for conditions (such as hypertension or alcoholism) on a one-to-one basis. (2) Company employees have an opportunity to enroll in small group sessions that use behavioral techniques and counseling to meet their interests and needs, such as smoking cessation or weight reduction. (3) Informational programs, which are generally targeted for the entire work force or large segments of it, are sponsored using outside experts or Bell System health professionals.

Cost

Studies of cost-effectiveness and cost benefits are being initiated. The New York Telephone Company is commited to what it calls a health care management program as a way of improving the health of its employees and containing the associated health costs. Experience to date has shown apparent significant reductions in the costs of absenteeism and hospitalization as well as other health-related costs, but it is still too early for specific quantification. Results from Bell System EAPs are also beginning to demonstrate cost-effectiveness.

Evaluation

Evaluation plays a key role in all Bell System health promotion efforts. Standardized education and evaluation protocols have been developed for smoking cessation, hypertension control, and detection and treatment of breast cancer. Computerized data collection and tracking systems are a major component of the Hypertensive Education Program and the EAPs.

Special Features

- Corporate Health Education and Employee Assistance managers charged with providing guidance and consultation on program development, implementation, and evaluation.
- Combination of selective targeted programs and broad-based programs.
- Focus on evaluation results in all program activities.

6

Anheuser-Busch Companies, Inc.

Sponsorship

The Employee Assistance Program (EAP) was developed jointly by management and 11 unions representing St. Louis workers. The United Labor Committee's Member Assistance Program, a Missouri labor organization representing the AFL-CIO, UAW, and Teamsters, assisted both management and Anheuser-Busch unions in program development and training. Since the program's inception in May 1978, company management and union representatives have met periodically to review the EAP for compliance with stated policies and procedures. A program booklet is distributed to all employees at a given site when the EAP begins, accompanied by a cover letter from the company president that affirms strong and continued top management support.

Objectives

The EAP was established to provide counseling for employees and family members on a wide range of personal problems, including alcoholism, drug abuse, emotional distress, marriage and family problems, and financial difficulties. It is designed to aid eligible persons suffering from the types of problems that may affect personal well-being and job performance.

Used with the permission of Anheuser-Busch Company, Inc.

Target Population

The program is designed for Anheuser-Busch and its subsidiaries throughout the United States and includes both employees and their immediate family members (18,000 employees). As of December 1980, the program was implemented at 23 sites. For the first 31 months of operation 1,590 persons used the program (74 percent employees, 26 percent family members). The annual penetration rate, corporate-wide, is 6.01 percent. For the first 31 months of operation, penetration rates varied from 3.5 to 19.4 percent at various sites. Utilization figures revealed that employees seek help in very close proportion to their prevalence (by demographic characteristics such as age range, job categories, etc.) on the work force.

Program Structure

The EAP is staffed by a team of counselors and limited support personnel. Service sites at St. Louis, Newark, Chicago, Los Angeles, and Fairfield, California, are staffed by Anheuser-Busch counselors. Services at the remaining 18 sites are provided through contracts with professional persons or agencies providing EAP services. Anheuser-Busch employees and members of outside firms are chosen on the basis of their understanding and warmth as well as past experience. A self-scored quiz is given at the end of a 2½-hour program orientation session attended by managers and shop stewards to see if they understand the program's purpose and procedures.

The key to the relatively high employee penetration rates has been identified as the emphasis placed on confidentiality. All program offices are housed in locations away from company premises. In addition, appointments are scheduled far enough apart that employees do not run into other employees seeking assistance. Because of the special handling of insurance forms, company personnel are unaware of the treatment an individual receives.

For the first 31 months, 81 percent of those using the program were self-referred (for alcoholism only—74 percent self-referred).

Program Processes

A person generally enters the EAP through one of four channels—self-referral, management referral, union referral, or medical referral. Supervisors may suggest that an employee seek help when there is a noticeable decline in his or her work performance that is not amended through usual supervisory procedures. At all points during the counseling process, the employee chooses whether or not to follow a recommended course of action.

Once the nature of the problem has been determined and a course of action is agreed on by both the counselor and the employee, the staff member may refer the client to another agency, medical professional, or local treatment center. In the event of referral (which occurs in most cases), the EAP counselor follows up to assure that the employee receives first-rate assistance and care.

The program is widely publicized in the hope that employees will decide to seek assistance before problems begin to seriously disrupt their work performance. Program brochures discussing family problems, alcohol and drug abuse, fear, stress and depression, and other subjects are mailed to the employees' homes several times a year. These materials outline the common sources of problems, describe the help available through the EAP, and instruct the employee or family member on how to arrange for an appointment. In addition, company publications are used periodically to inform employees and family members about the program.

In June 1981 the company participated in Duke University's Annual Summer Institute of Alcohol Studies by offering an Employee Assistance course for alcohol counselors. The course was four and a half days long and explored ways to develop and maintain an employee assistance counseling program.

Cost

Total annual costs incurred by the company to operate the EAP nationally are under the mid range in a six-figure budget.

Initial counseling services are offered free of charge to participants. The company's group health insurance program was broadened when the program started to include both inpatient and outpatient coverage for alcoholism and drug abuse as well as increased psychiatric and psychological benefits.

Evaluation

In the interest of confidentiality, program data are processed by an outside firm—in spite of in-house capability. Program managers are convinced that the computerized information system being used by personnel at 22 sites is highly accurate and generates timely and valuable data. During 1982, the company intends to use the system and an additional phase of programming to provide cost comparisons on group health insurance and sick leave payments both before and after treatment.

Special Features

- Periodic mailing of descriptive program brochures to employees' homes.
- Data processing by an outside firm in the interest of confidentiality.

7

Blue Cross and Blue Shield of Indiana

Sponsorship

The Health Promotion Service was developed by the American Health Foundation, a scientific research organization with a strong interest in preventive health care. The Blue Cross/Blue Shield management strongly endorses the program.

Objectives

The Health Promotion Service of Blue Cross/Blue Shield of Indiana uses this health risk identification and reduction program because it is targeted for the worksite and enlists organizational employees to help deliver services. The program focuses on the three leading causes of mortality in the United States—cardiopulmonary diseases, cancer, and cerebrovascular disease. Peer group support is used as a means of encouraging participants to make the changes necessary to improve not only their current health status but also their future health prospects.

Used with the permission of Blue Cross and Blue Shield of Indiana.

Target Population

Blue Cross/Blue Shield of Indiana has 2,486 employees who are eligible to participate in the program. Of that number, slightly more than 80 percent fill out the initial questionnaires and continue through the program (as of early 1981, approximately three years after its inception).

Program Structure

The core staff consists of a team of three "health interventionists." The interventionists are required to have a health education background, group dynamics experience, strong writing and speaking skills, and good interpersonal relations.

Three basic concerns are reflected in the program: management cooperation, voluntary participation, and record confidentiality. In an effort to encourage program participation, an active educational and promotional campaign involving slide presentations, large art displays, posters, brochures, and electronic media messages has been directed to employees. A periodic newsletter discusses program design and participation as well as such topics as improving dietary habits. These numerous sources are designed to inform employees about the concept of risk and to affirm that their health may be enhanced by life-style modification.

Program Processes

Eligible employees who are interested in the program receive an appointment for a physical mini-exam (height, weight, cholesterol and carbon monoxide levels, and blood pressure) from their supervisors. Meanwhile, they fill out a confidential questionnaire that, combined with the information provided during the brief physical, provides the basis for individualized health hazard appraisals. Exit interviews are arranged between health counselors and program participants to discuss general health status, and high-risk individuals are encouraged to enroll in the intervention program. Appropriate medical referrals are made. The results of the mini-exam are recorded in a personal "health passport," which allows the participant to keep a five-year record of such variables as blood pressure and alcohol consumption.

Following the two-part examination and data processing, an individualized letter regarding risk is mailed to the participant's home, along with a pamphlet describing ideal values and their relationship to norms. In each letter, specific risk factors are discussed, with a recommendation about how to reduce them. Participants have a second opportunity to enroll in a risk intervention program after receiving their letters.

The intervention program consists of classes in nutrition education, smoking cessation, weight reduction, stress management, drug/alcohol abuse, and fitness—all sponsored by the Health Promotion Services. Group sessions last about one hour and are held once a week for 6 to 16 weeks, depending on the particular group. The groups are goal-oriented and behavior modification techniques are emphasized. Participants are encouraged to develop their own strategies for changing their life-style. Those who complete a particular portion of the intervention program or who were referred for medical care are telephoned periodically to monitor their progress. At that time, they are encouraged to continue shaping their new behavior, and reminded that other program participants need their support and encouragement as well.

Cost

The total per capita costs amount to $30.00 per year, while the base screening cost itself is roughly $7 to $10 for each employee.

Evaluation

Program data are collected and stored on computer file for comparative analysis, long-term follow-up, and ongoing program evaluation. One of the key objectives of the follow-up and maintenance (final) program phase is to gather the data necessary to make long-run cost/benefit calculations. Program and self-evaluation forms are completed by the participant after the screening and intervention process. The results of this inquiry help program staffers with their evaluation and program planning responsibilities.

Special Features

- Program developed by outside research organization.
- Active educational and promotional campaign.
- "Health passport" encourages long-term commitment to health promotion.
- Strong emphasis on program evaluation.

8

Campbell Soup Company

Sponsorship

The programs have strong management sponsorship.

Objectives

Although primarily aimed at early disease detection, the periodic health examination has a follow-up component designed to allow individual counseling about health risk avoidance.

Target Population

Most programs are offered at Campbell plants nationwide, and there are roughly 30,000 employees eligible to participate in them.

Program Structure

The company sponsors various health promotion and disease prevention initiatives involving hypertension screening, atherosclerosis prevention, breast and rectal cancer detection and prevention, weight reduction, serum lipid modification, and exercise programs. Campbell's atherosclerosis prevention program is largely managed by nurses with minimal physician su-

Used with the permission of Campbell Soup Company, Inc.

pervision. Long-term follow-up and intervention opportunities are among the program features. Physician referral for outside assistance is provided where consistently high blood pressure is indicated. Several plants have used a group approach for their behavior modification programs.

Program Processes

There has been a major effort at Campbell in recent years to design high visibility campaigns and programs. Posters and other forms of communication are used regularly to promote such activities as hypertension reduction. Recent surveys indicate that employee awareness of the presence of certain conditions such as hypertension has increased dramatically.

In addition to the programs listed in the previous section, the company experimented with a Family Health Carnival for employees and their families in 1978. Approximately 1,400 people attended and follow-up questionnaire results indicated that the information provided to many carnival participants influenced their safety and health behaviors.

Cost

Recognizing that the benefits of a program such as hypertension control necessarily accrue in the long run, corporate sponsors do not expect the upfront money spent to be returned immediately. The annual cost of in-house diagnostic work and treatment at four Campbell locations was estimated to be roughly $100 per patient.

A behavior modification program for 70 volunteer smokers cost $500 for each person who was a nonsmoker one year after the completion of his/her group program. The company believes that the 20 percent success rate might be higher if the $20 participation fee were raised, although they fear that this would reduce participation.

The company recommends a mammography at least every four years for women between the ages of 35 and 50 and at least every two years thereafter on the assumption that such a policy is cost-effective.

Evaluation

For 1978, data review was primarily aimed at evaluating the degree to which risk factors associated with smoking, hypertension, elevated serum cholesterol, elevated serum triglycerides, and obesity had changed over a given period. Surveys are conducted periodically to gauge employee awareness of certain conditions and to monitor the impact of various health promotion/ disease prevention strategies.

Special Features

- Major program promotional effort.
- Experimentation with one-shot programs such as the Family Health Carnival.
- Minimal physician supervision for programs that are nonetheless largely staffed by medical personnel.

9

Ford Motor Company

Sponsorship

Management is the primary sponsor of the program, with union and community service involvement in selected program components.

Objectives

The overall objective is health promotion. This is approached through a variety of programs aimed at risk intervention and life-style modification.

Target Population

Target populations vary with the several categorical and specialty programs. For example, 20,000 employees (both blue- and white-collar) are eligible for a pilot hypertension Control Program, while 5,000 salaried (white-collar) employees are eligible for the Cardiovascular Risk Intervention Program (CRIP).

Program Structure

The program's several elements are differently structured. All program participation is voluntary and the results are strictly confidential. The in-house program elements include:

Used with the permission of Ford Motor Company, Inc.

- Preemployment physicals for all new employees.
- CRIP—includes screening and risk reduction (see description in the next section).
- Smoking Cessation Program—a spin-off from the CRIP. It includes a self-directed program booklet consisting of daily assignments that are provided to persons who want to quit or cut down on smoking.
- Hypertension Control Program—An NHLBI (National Heart, Lung and Blood Institute) grant for a three-year demonstration in four plants. UAW cosponsors this program, as blue-collar employees are included.
- Periodic special programs—e.g., hypertension screening for everyone in conjunction with a Blue Cross/Blue Shield "Fair."

The "out-house" programs are conducted in conjunction with the Southeast Michigan Chapter of the American Cancer Society and feature:

- Breast cancer education program and some examinations at four locations.
- Colo-rectal education and screening (hemocult testing) at two locations.

The corporate headquarters states general policy on health promotion, but has very little authority over plants.

Program Processes

The CRIP includes the following components:

- Recruitment for volunteers through general in-house information channels.
- Volunteers screened by nurses in the Medical Department to determine their risk of developing premature coronary heart disease. Cholesterol, blood pressure, and weight readings are combined with information regarding family history and personal life-style to pinpoint the causes of a possible risk.
- Risk assessments made; all smokers referred to the Smoking Cessation Program, and all classified into one of five risk categories ranging from low to high. Those with one or more risk factors are invited to attend intervention sessions (a special effort is made to assure attendance by the higher-risk individuals); the very high-risk types are referred to physicians.
- Risk intervention sessions held after work on employee's own time; five

one-hour sessions in a six-week period; aimed at life-style changes. Employees who participate in the CRIP are given a small handbook that discusses factors related to heart disease and describes the steps employees can take to mitigate their impact or eliminate them.

- Annual rescreening and reattendance at intervention sessions if indicated.

This program also includes a Heart Healthful Menu component for in-house cafeterias. Originally available only in World Wide Headquarters, it has now been extended to 14 locations in the Detroit area. A walking program, referred to as the "headquarters hustle," is another component of this risk intervention procedure. Interested employees are given a pamphlet that outlines possible routes to walk and includes the distances covered. Prospective participants are given instruction about how often and how vigorously to exercise.

The Hypertension Control Program, a demonstration program, is conducted at four plants and includes both blue- and white-collar employees. It is voluntary and on the employee's own time. This demonstration includes screening, referral, and follow-up. It has a control group plus three intervention models: One is on-site treatment (Alderman model) and the other two are variations of referral and follow-up. Scheduled to be completed in June 1981, the demonstration has been extended for an additional eight months.

Cost

Data on the costs of the CRIP are not available. Such information will probably await some cost-benefit results. The annual total cost of the Hypertension Control Program is roughly $300,000.

Evaluation

The company's proposed evaluation of the CRIP is designed to yield estimates of both direct and indirect expenses. A subset of Ford Company nonclients would be matched demographically and compared with a set of program participants. Ford is particularly interested in deriving measures of the net benefit as well as the benefit-cost ratio of the CRIP.

As of January 1980, of the 1,722 screens performed in the CRIP, 44 percent were classified moderate or high-risk; 48 percent were overweight; 33 percent had elevated cholesterol levels; 32 percent were smokers; and 12 percent had elevated blood pressure.

Special Features

- Program scope varies widely among locations—highly decentralized operation.
- Target population for certain program components is specified on the basis of risk factor prevalence.

10

IBM Corporation

Sponsorship

IBM's health care strategy is built on management support as well as close cooperation with community resources (YMCA, American Cancer Society, American Heart Association, etc.). Occupational health and employee drug-abuse programs began during the 1950s, multiphasic screening in 1968. Major health promotion activities, including the health education program called A Plan for Life, were launched in February 1981.

Objectives

IBM's program has three main objectives: to encourage individual responsibility for health; to offer a series of voluntary health education and assessment programs to heighten awareness of individual health status and needs; and to encourage employees to become actively involved with hospital and health maintenance organization (HMO) boards, health systems agencies (HSAs), and nonprofit organizations like the American Cancer Society and the American Heart Association.

Target Population

Employees, retirees, and members of their families are eligible to enroll in the health education courses. The more traditional occupational health and

---

Used with the permission of IBM Corporation.

multiphasic screening programs are offered to IBM employees throughout the country. (Voluntary Health Screening is offered to employees 35 years of age and older.) Since the beginning of the screening program more than 10 years ago, there have been approximately 190,000 voluntary health exams conducted. Through these exams 41 percent of the population was found to have unsuspected medical conditions. Occupational health programs are required on a medical surveillance basis.

Program Structure

There are seven elements to the IBM health care strategy: (1) health education sessions for employees and their families and for retirees and their spouses; (2) emphasis on early detection and prevention rather than treatment; (3) a wide range of community-oriented employee programs to encourage involvement at the local level; (4) development of measures for cost-effective benefits; (5) contribution to local health facilities and organizations and development of HMOs; (6) contributions to and staff involvement in health research; and (7) participation in various health organizations on both the national and local levels.

The company employs an extensive medical staff, including 50 physicians and more than 150 nurses at major facilities in the United States. The health education program, A Plan for Life, includes courses in exercise, smoking cessation, stress and weight management, alleviation and prevention of back problems, first aid, cardiopulmonary resuscitation, nutrition, and risk factor management. These courses are available free of charge when scheduled by IBM, and on a tuition-assistance basis when conducted by outside organizations for the general public.

Program Processes

Every month an IBM computer identifies the employees who are eligible to participate in the company's screening program. These persons receive an invitational letter and descriptive brochures, plus a personal medical history questionnaire. Participating employees complete the questionnaire and make an appointment through the medical department for the screening exams. Company nurses conduct the screening exam, which includes blood pressure, height and weight, visual acuity and hearing tests, chest X ray if clinically indicated, and drawing of a blood sample for various tests. A hemocult slide is provided for colo-rectal cancer tests. Physicians are not routinely involved in the exam, and employees are encouraged to consult their personal physicians for follow-up once the results are processed and made available. IBM's

A Plan for Life program offers a variety of health education programs that complement the screening exam. Employees are encouraged to participate in such courses whether or not they have recently completed a voluntary screening exam.

Cost

The company's strong commitment to cost-effectiveness is clearly reflected in program structure. Family members and retirees are eligible to participate in the health education program since they contribute two-thirds of the company's total outlay for health care. IBM emphasizes the need for individual responsibility but assumes that its company-sponsored preventive health program will be cost-effective in the long run. Finally, the company draws heavily on community resources and facilities, believing this to be a sound approach for containing program costs.

Evaluation

An April 1980 report showed that the multiphasic screening program initiated in 1968 had yielded the following results: 26 percent of participating employees were found to be "healthy"; 33 percent had a condition that they knew about before the screening; and 41 percent required further attention for conditions (many of which were minor) that they were previously unaware of.

The first offering of A Plan for Life courses in spring 1981 resulted in very positive attitudes and high participation. Over 1,200 mini-courses and comprehensive courses were conducted, with total enrollment exceeding 25,000. In addition, about 1,300 people attended courses under tuition assistance and 22,000 requests were received for the American Lung Association smoking-cessation manuals. Enrollment requests broke down as follows: 68 percent employees; 28 percent employee spouses; and 4 percent employee dependent children, and retirees and their spouses. The most popular courses were exercise, stress management, weight management, and cardiopulmonary resuscitation.

Special Features

- Availability of both short and more comprehensive courses on health promotion and disease prevention, as well as tuition assistance for eligible community programs.

- Inclusion of spouses, teenage children, and retirees and their spouses in health promotion programs. No distinction between management and nonmanagement employees.
- Heavy reliance on community resources and facilities.
- Employees strongly encouraged to serve on outside health boards and agencies and to assist in current health research (company pays for a certain amount of these activities).
- Company also has extensive automated occupational health monitoring program for protection in the workplace.

11

Internorth, Inc. (Northern Natural Gas)

Sponsorship

Although the program is endorsed by management, it is supported mainly by membership dues. Community representatives assist with cardiopulmonary resuscitation classes, blood pressure screening, and recreational activities.

Objectives

The overall goal is to promote an awareness of healthy life-styles among employees. Specifically, the cardiac rehabilitation program is designed to educate employees with or at high risk of heart disease about ways to reduce the probability of heart attack. In addition, the fitness center was developed to provide employees with enjoyable and healthful recreational activities while encouraging an appreciation of physical fitness.

Target Population

Various program activities are generally limited to company employees, although the entire family is eligible for fitness center offerings. There are approximately 900 fitness center members from a total eligible employee pool of nearly 1,700.

A release waiver is signed by each member during the initial sign-up procedure. For those persons 35 or older and for those under 35 with two or more coronary risk factors, medical clearance is required before they are eligible to partake in the fitness program.

Program Structure

An extensive fitness center offers a site for stationary cycling, basketball, swimming, healthy-back classes, weight training, racquetball and squash, jogging, and aerobic dance. The center's staff includes a director, health/fitness specialist, activity coordinator, and several part-time assistants. The director and fitness specialist are qualified to administer and evaluate the fitness tests and to prescribe exercise programs.

The center staff offers a thorough testing program, and fitness specialists develop personalized exercise regimens based on test results. Fitness evaluations are designed to measure aerobic capacity, flexibility, body composition, muscular strength, and endurance. Exercise sessions engage participants in 30 minutes of aerobic activity at a prescribed exercise heart rate. Warm-up and cool-down exercises, along with weight training, are used to supplement the aerobic phase of training. Each participant is supervised by qualified exercise leaders who are trained in cardiac rehabilitation. Fitness testing results are held in strict confidence by the program's staff.

A series of behavior modification sessions is available to employees and serves as an adjunct to the exercise program in an effort to promote healthful life-styles. Health education classes are designed to instruct members about the importance of physical fitness, weight management, and control of heart disease risk factors. Nutrition and smoking cessation classes are also offered on a continuous basis. Outside specialists are occasionally enlisted to conduct stress management workshops.

A quarterly catalogue announces the fitness center's class offerings (with a particular emphasis on new classes) and briefly discusses course content. The newsletter "Pulse" comprises a section of this catalogue and focuses on diet/nutrition, stress management, and exercise.

Program Processes

A computerized fitness profile is constructed on the basis of test results and individualized fitness prescriptions are designed to reflect these profiles. Employees are periodically retested to compare changes in fitness levels with measurements derived at other times throughout the year.

Cost

Membership dues supplement the funding provided by the company.

Evaluation

Each participant in the fitness center program is trained to take his or her own pulse rate and to compare readings at different stages of exercise. Periodic laboratory evaluations of pulmonary function, percentage of body fat, blood pressure, and physical work capacity are made and the data derived are provided to private physicians on request.

Special Features

- Strong emphasis on the importance of shaping new behaviors in the long run—for example, through periodic retesting of participants and continuous exercise logs kept current by employees.
- Program data shared with private physicians.
- Program open to all levels of employees.
- Cardiac rehabilitation program.
- Complete fitness lab designed to evaluate:

  —Pulmonary function

  —Aerobic capacity (with stress electrocardiogram)

  —Body composition (with the use of calipers or hydrostatic weighing technique)

  —Muscular strength, endurance, and flexibility

  —Personalized exercise prescriptions based on individual fitness levels

  —Nutritional advice

12

Johnson & Johnson, Inc.

Sponsorship

The Johnson & Johnson Live for Life program is a health promotion effort intended ultimately for all Johnson & Johnson employees worldwide. The Live for Life program is specifically designed to encourage employees to follow life-styles that will result in good health. The program is based upon these assumptions:

1. Life-style activities such as eating, exercise, smoking, and stress management contribute substantially to an individual's health status.
2. Life-style activities that support good health can be successfully promoted at the work setting.

Objectives

During the initial three-year phase, Johnson & Johnson is committed to a careful evaluation of the Live for Life program in regard to its impact on employee health and its overall cost-benefit to the corporation.

The Live for Life program began with two primary goals:

1. To provide the means for Johnson & Johnson employees to become among the healthiest employees in the world
2. To determine the degree to which the program is cost-effective

---

Program objectives include improvements in fitness, nutrition, weight control, stress management, smoking cessation, and health knowledge. The proper utilization of such medical interventions as high blood pressure control and the Employee Assistance Program is strongly encouraged. It is anticipated that such improvements will lead to positive changes in employee morale, relations with fellow employees, company perception, and job satisfaction and productivity, as well as reductions in absenteeism, accidents, medical claims, and total medical costs.

Target Population

Responsibility for the program rests with Live for Life staff at the corporate level. As of July 1981, this staff serves about 8,000 employees in active programs. Live for Life is primarily a service organization. Its mission is to provide the direction and resources to Johnson & Johnson employees, their families, and the community that will result in healthier life-styles and help contain illness care costs. It supplies participating companies with the consulting expertise, training, core program components, professional services, and promotional materials necessary for a program's success. The Live for Life staff is also responsible for program development and evaluation.

Program Structure and Processes

It is important to recognize that Johnson & Johnson is a highly decentralized group of companies, each of which operates in a very independent fashion. Acceptance of and full commitment to the program by a company's senior management is essential to ensure the financial backing and time allotment to support the Program.

Following a decision by a company management to accept and support the program, voluntary employee leaders at that company are selected and trained to manage it. Working closely with Live for Life staff, these employee leaders assume primary responsibility for promoting good health practices at work among their fellow employees. Employee participation in the program is voluntary and free of charge. With a few exceptions, employee participation is off company time. Development of exercise facilities, improvement of the quality of food in the company cafeteria and establishment of a company smoking policy are examples of worksite environmental changes undertaken by program leaders. The employee leaders also plan the introduction of individual life-style improvement programs and group activities for all employees. Most of these programs are initially supplied through Live for Life. They include a Health Screen, which allows employees to examine how healthy their current life-styles are; a Lifestyle Seminar, which introduces employees to the Live for Life concept in depth; and a variety of life-style improvement programs in such areas as exercise, smoking cessation, stress management, nutrition, weight control, and general health knowledge. Live for Life activities are integrated closely with established medical programs (e.g., high blood pressure detection and control), and Employee Assistance Programs.

Evaluation

A two-year epidemiological study is in progress to evaluate the impact of the program on a wide range of employee health and life-style characteristics. These variables, which are collected annually, include the following measures: *biometric* (e.g., blood lipids, blood pressure, body fat, weight, and estimated maximum oxygen uptake), *behavioral* (e.g., smoking, alcohol use, physical activity, nutrition, healthy heart behavior pattern, job performance, and human relations) and *attitudinal* (e.g., general well-being, job satisfaction, company perception, and health attitudes). Using a quasi-experimental design, four companies receive the complete Live for Life program, while five companies participate as "controls," offering only the Health Screen to their employees annually. Health and life-style information is collected through the Health Screen at all epidemiological sites at baseline and at the end of years one and two.

Approximately 4,000 employees are involved in the epidemiological study sample, almost equally divided between control and active sites. The epidemiological study should be complete in fall 1982. At that point, the full Live for Life program will be introduced at the "control" sites, thus allowing a replication of any intervention effects.

Costs

Preliminary work is also under way to measure the cost-benefits of the program. Since Johnson & Johnson is self-insured for illness care costs, any changes in the number or dollar amount of illness care claims attributable to a positive health program are of considerable interest as a measure of the program's benefits. Other potential benefit measures include absenteeism, turnover rates, accident rates, and a host of employee and management attitudes toward themselves, their work, and one another. It is felt that this is pioneering work in an area where potential benefits have been difficult to measure with existing systems and methods.

The program has, from its inception, been a multidisciplinary effort involving professional assistance from a variety of scientific, academic, and commercial institutions. Although behavioral scientists have played a key role, many medical disciplines have become involved from the start, as have epidemiologists, health educators, and health economists. However, in the final analysis, it is the managers and employees of Johnson & Johnson companies and the Live for Life staff who have carried the major responsibility for the program's implementation and modification and for the success that has been achieved.

Special Features

- The health screen accentuates positive health rather than health risk.
- Resource center provides a centralized source of information and assistance for instructors and participants alike.

13

Kimberly-Clark Corporation

Sponsorship

The health management program was initiated by management during June 1977.

Objectives

The main goal of the program is to achieve a higher level of wellness among employees and to thereby enhance productivity while reducing absenteeism and health care costs. Once participants are informed of their own health risks, they are encouraged to make changes in their life-style—a life-style that might predispose them to illness if left unchanged.

Target Population

All of Kimberly-Clark's salaried employees in the Fox Valley area of Wisconsin have an opportunity to participate voluntarily and without charge. Of the 4,500 eligible employees, 4,250 had taken part in multiphasic screening and exercise tests as of February 1, 1981. Participation in the exercise program currently is 25 percent of the eligible screened employees.

Used with the permission of Kimberly-Clark Corporation.

Program Structure

A $2.5 million multiphasic testing and physical fitness complex was constructed to serve as a base for the health management program. The facility has a 100-meter walking and running track, an Olympic-size pool, exercise equipment, a sauna and whirlpool, as well as locker and shower areas.

The principal program components consist of a medical history and health hazard appraisal, multiphasic screening, physical examination, exercise test, and a health review with recommendations. An aerobic exercise program offers jogging, swimming, stationary cycling, circuit training, aerobic dancing, rope jumping, water exercise, cardiac rehabilitation, and cross-country skiing. Special exercise classes are available for those who have had heart disease or emphysema. Health education classes cover a broad spectrum of topics, including diet management, smoking control, breast self-examination, high blood pressure, and cardiopulmonary resuscitation. Individual counseling is also available. The Employee Assistance Program is available to help employees and their families with drug and alcohol dependency and other special health problems.

The total program staff consists of 25 full- and part-time personnel, including medical personnel, exercise specialists, life guards, and secretarial/receptionist staff. A physician, registered nurse, and technician routinely attend the supervised cardiac rehabilitation exercise sessions. Community participants include a physician advisory council, family practitioners and internists who help with exercise testing and physical examinations, and consulting cardiologists, radiologists, and dietitians.

Program Processes

The program consists of a series of activities designed to familiarize employees with the characteristics of a healthy life-style and to encourage participants to adopt these habits as part of their daily routine. A worker who wishes to begin the program completes an extensive medical history. Next, the participant is given a battery of laboratory tests to assess such conditions as blood pressure and content, organ function, and body structure.

A complete physical examination, including a treadmill or bicycle exercise test, is conducted by Health Services Center professionals. The program staffers then review test and exam results with each employee and recommend certain health promotion measures, such as dietary counseling and exercise. After their health review, employees may use the physical fitness center to perform their prescribed exercises.

More than 30 letters, brochures, and newsletters have been used to communicate information concerning the health management program to employees. Twelve quarterly issues of a newsletter entitled "Wellness" have been distributed since March 1978. They describe upcoming classes, health and fitness tips, and program results. A travel diary with pointers on keeping fit while on the road has also been given to employees.

Cost

The annual cost of the program, which runs approximately $260 per capita including complete medical exam, is entirely covered by the company.

Evaluation

A monthly statistical summary includes participation figures, demographic characteristics of program participants, and information about employees who have received follow-up attention. To gauge the effectiveness of the plan, the firm intends to use computerized medical histories to document changes in employee health status. Kimberly-Clark is also working with its insurance carrier to compare the incidence of illness and costs of hospitalization, using program participants and a control group.

Various reports have highlighted some interesting findings in regard to the program. There has been an extremely low positive yield from chest X ray and cancer screening—just a handful of cases detected from among several thousand employees who were checked. The number of visits to the Neenah facility's dispensary nurses for minor illnesses were less during 1979 than in 1978 and remained lower in 1980. Although there is no compelling evidence that the health management program is responsible for this trend, it is noteworthy that no similar decrease was found in certain other Kimberly-Clark units lacking the program. The company's preliminary overall results do suggest that a program of regular physical exercise significantly affects a subjective sense of well-being and physical work capacity.

Special Features

- Fairly rigorous data collection and analysis, especially by using Kimberly-Clark employees without the benefit of a program as a control group.
- Extensive information campaign designed to familiarize employees with program content and availability.

14

Metropolitan Life Insurance Company

Sponsorship

Metropolitan Life Insurance Company's Health and Safety Education efforts span almost 75 years of active community involvement. More recently, its scope has broadened to encompass occupational health promotion programs. In addition to establishing its own health promotion program for home office employees—the Center for Health Help—Metropolitan offers consultation services and educational materials to assist its group insurance customers in developing and tailoring their own programs.

Metropolitan Life Insurance Company consists of a home office in New York City and nine head offices throughout the United States and Canada. Each company unit independently administers health promotion programs on site through its Medical Department.

Objective

The primary objective of the Center for Health Help is to develop, provide, and evaluate cost-effective health promotion programs that place a priority value on life-styles that positively influence health.

Target Population

The Center for Health Help's services are available to the total employee population of the company, and the expertise of its staff is offered on a consultant basis to its group policyholders. Specifically, the center focuses on prevention, although it also provides selected services for those employees with existing health concerns or conditions, such as diabetes and hypertension.

Program Structure

Functioning as independent but interrelated units, Metropolitan's comprehensive employee health program includes medical services through its Employee Health Conservation Unit; counseling for alcohol/drug abuse and for emotional, financial, and preretirement problems through its Employee Advisory Services; exercise opportunities through its Employee Activities Unit; and prevention services through its Center for Health Help.

The Center for Health Help offers individual counseling and group programs on topics like smoking cessation, nutrition education, weight control, cholesterol reduction, breast cancer education, and stress management. All programs are free of charge to employees, but are held on personal time, except for the cholesterol reduction program, for which two hours of company time is allowed.

Cost

The annual cost of providing the Center for Health Help's programs is around $15 to $20 per eligible employee at the home office in 1981. Specific topical units are being evaluated to determine their long-range cost-effectiveness.

Special Features

A stratified, random sample consisting of 20 percent of the home office population was surveyed in 1980 to provide benchmark data on employees' health status, history, attitudes, practices, and interests. The results of the computer analysis of this survey will be used to further identify and meet health needs of the employees.

Evaluation of the smoking cessation and cholesterol reduction modules will be completed and presented at professional conferences and in the literature beginning in late 1981.

15

National Aeronautics and Space Administration

Sponsorship

The extensive health prevention and disease prevention activities begun in 1962 are sponsored by management.

Target Population

All NASA employees at the Ames Research Center are eligible to participate in program activities, and family members are allowed to attend some of the nutrition sessions. Seventy percent of the nearly 2,000 eligible workers have participated, and the average age of participants is 46.

Program Structure

There are numerous components of the health promotion and disease prevention program at the NASA-Ames facility, including hypertension screening, stress electrocardiogram, multiphasic health screening, employee assistance programs, stress management, weight and nutrition classes, smoking cessation, and fitness.

A program of annual health examinations was recently expanded to include counseling based on a computerized appraisal of specific health risk factors. Each interested employee may have an individualized health hazard appraisal consisting of a computer analysis of the answers to a six-page questionnaire concerning such topics as medical history and life-style. Health unit staffers take some of the required physical measurements, although a reporting system involving employee initials and identification numbers has been used to ensure privacy. The appraisal displays a number of weighted risk factors and illustrates their relation to the 10 leading causes of death as determined for that patient. Health hazard assignments at Ames are designed to supplement rather than to replace the periodic health examinations.

Program Processes

The patient makes a counseling appointment for hazard appraisal shortly after the health examination. A physician explains test results and discusses any risk factors that deviate significantly from the norm and thereby constitute a health risk. By comparing the patient's true age with his/her risk age as determined by the appraisal, the advising physician may recommend various life-style changes. In addition, the participant may be referred to his personal physician for required treatment.

Cost

A 1975 cost analysis of the multiphasic health testing program at the Ames Center concluded that unit costs for such a low-volume operation (then 30 patients per week) are similar to those reported for automated programs that accommodate as many as 500 patients during the same period. Cost comparability appeared to have resulted from personnel savings and smaller space requirements for the Ames program as compared to larger programs.

Program costs currently run approximately $200/person/year to maintain services. Seventy percent of the total budget is tied directly to promotion and prevention activities.

Evaluation

NASA representatives have concluded that health hazard appraisal counseling is an effective means of altering individual health practices. The first annual retesting of a group of more than 100 examinees yielded a net risk reduction age of 1.4 years, while a longer term follow-up showed a reduction of more than 2 years in a smaller group of patients. The net risk reduction ages in the two groups represented 32 to 40 percent, respectively, of the achievable risk reduction ages when patients comply with recommendations made during counseling sessions.

The patients' acceptance of the appraisal seems consistently high. A recent survey revealed that four-fifths of the participants strongly endorse the program and wish to be retested annually. Fifty percent interpreted their test results as pointing to necessary health practice changes, and eighty percent indicated that they intended to make some or all of the necessary modifications to lessen the risk of death in the next 10 years.

Special Feature

- Computerized hazard appraisal used as an adjunct rather than as an alternative to annual health examinations.

16

Sentry Life Insurance Company

Sponsorship

Management has approved the program since its inception in 1977.

Objectives

Individualized fitness programs are designed to recognize each participant's level of fitness as well as his or her health goals. All programs are geared to keep exercise interesting while promoting overall fitness.

Target Population

Sentry employees and members of their immediate families may use the physical fitness center regularly. The health education classes as well are not targeted to employee subgroups.

Program Structure

The fitness center offers a variety of facilities, including a 25-meter pool, full-sized gymnasium, racquetball and handball courts, an indoor-golf driving

---

Used with the permission of Sentry Life Insurance Company.

range, cross-country skiing trails, and a vast array of weight-training equipment, along with a cardiovascular fitness laboratory.

Several unique features characterize the Sentry health promotion program. A "quiet room" specifically designed to allow employees to seek refuge from the work environment has been designated, and brochures encourage employees to take relaxation breaks rather than coffee breaks. (The printed material also discusses stress reduction methods and relaxation techniques.) Special offerings have included glaucoma screening, hypertension screening, life-saving techniques, a cancer education day, low-back clinics, a class on healthful cooking, stress management programs, adaptive fitness programs, alcohol awareness programs, dental awareness programs, aerobic dance classes, and classes in cardiopulmonary resuscitation.

Health professionals conduct weight control programs at regular intervals and provide general information on dietary principles. They also periodically hold stop-smoking clinics that are designed to encourage individuals to abstain from smoking and to prod those employees who have relapsed to consider a second-best alternative, periods of nonsmoking.

Fitness training, aquatic fitness, slimnastics, and tournaments are just a few examples of the programs offered within the fitness center. Other activities include cross-country skiing, jogging, first aid, personal defense, swimnastics, and conditioning for long-time fitness. The medical staff performs both at-rest and stress testing as a measure of cardiovascular fitness for employees who engage in fitness center activities. They encourage aerobic exercise programs as the ones that most significantly improve blood flow and lung capacity.

A diverse staff consisting of medical personnel and health education personnel as well as physical fitness specialists is responsible for conducting the numerous Sentry health promotion activities.

Program Processes

A person using the fitness center for the first time is given an orientation by the staff for one hour, during which he or she receives an explanation of the company's philosophy of health (a holistic approach with a heavy emphasis on individual responsibility) and is apprised of the various options for physical fitness programs that are available. Prospective participants are given a protocol to fill out that gives background information for stress testing and includes questions on dietary habits and medications taken. A medical department nurse takes the employee's blood pressure and pulse, reviews the medical history, and decides whether or not further medical evaluation is needed. A physical fitness specialist helps each participant design a fitness program to suit his/her needs and interests.

Cost

The costs of the entire health promotion program are absorbed by the company.

Evaluation

There is a very limited data base presently available, but Sentry is in the process of expanding it. Program representatives are particularly interested in cost figures and absenteeism trends.

Special Features

- "Quiet room" provided as a relaxation area and employees are informed about appropriate stress reduction exercises they may wish to perform while using this area.
- Wide variety of course offerings within the fitness program, including instruction in first aid, personal defense, and cross-country skiing.
- Major emphasis on cardiovascular health, as affected by proper diet, exercise, blood pressure control, smoking cessation, and self-responsibility in risk reduction.
- A variety of recreational activities to facilitate employee interest.

17

Trans World Airlines, Inc.

Sponsorship

The TWA Special Health Services Program, which includes handling of substance abuse referrals, is jointly sponsored by management and labor. In the early 1970s the company's insurance carrier adopted alcoholism as a fully covered disease on a par with other illnesses. As the program developed, its base was broadened to include other disorders that may affect job performance.

Objectives

The main objective of the TWA Medical Department's Special Health Services is to provide assistance and offer treatment to employees suffering from behavior-medical disorders such as alcoholism, mental health problems, and drug abuse. The ultimate goal is to prevent these conditions from progressing to the point where those with these problems cannot work and live effectively. One of the primary program aims is to follow a holistic approach—one that is concerned with medical examination as well as consultation in regard to the problem for which the employee (or family member) was referred.

Target Population

The main target employee subgroups include pilots, flight attendants, mechanics and ramp service personnel, clerical staff, and management. The

Used with the permission of Trans World Airlines, Inc.

single-program approach is designed to be flexible enough to recognize special subgroup interests.

Program services are also available to the employee's immediate family members and dependents. A pool of approximately 96,000 persons (i.e., 32,000 employees multiplied by 3 dependents for each employee) is eligible for assistance.

Program Structure

TWA has designated four major criteria for the primary treatment programs it selects for its employees and their dependents: (1) maximum family involvement; (2) exposure to community support services and self-help groups such as Alcoholics Anonymous; (3) an overriding goal of having patients who are chemically free of drugs on discharge; and (4) a willingness on the part of the treatment staff to help establish and become involved in a long-term Continuing Recovery Program (CRP), which provides after-care assistance to patients.

The core program staff consists of program managers in Kansas City, New York, and Los Angeles who work with prediagnosis, intervention, and case monitoring; they are actively supported by the airline's staff vice president of Health and Environmental Services and his area medical directors. Program staffers do not consider themselves as being involved with primary treatment, but they become heavily involved with each case in the long-term CRP.

Program participants, who are either self-referred or approached by supervisory and/or union personnel, are placed in general community hospitals with accredited substance-abuse recovery programs or referred to free-standing facilities that specialize in the treatment of alcoholism and/or other drug dependencies. TWA program personnel expect treatment facilities to involve families in the recovery process as early as possible in the treatment cycle. All facilities on the referral list are regularly monitored to assure compliance with TWA treatment criteria.

Training sessions are currently under way for shop stewards as well as supervisory/managerial personnel on how to prevent rather than to react to substance abuse problems. However, trainers emphasize that only the program medical staff may actually diagnose the drug dependency problem as such.

Several major forms of communication are used to publicize program services. Articles announcing the program describe the specific disorders it covers; union newsletters and newspapers identify program personnel and discuss their functions; and letters mailed directly to the homes of select subgroup employees outline available services and steps to follow for interested employees and family members.

Program Processes

The initial interview following an individual's decision to seek help focuses on job performance. The employee is informed of services that are available on a strictly confidential basis. Any refusal to accept referral for diagnosis or to follow prescribed treatment is handled in accordance with contractual agreements between union and management.

Pilots who receive treatment are given a copy of a manual that outlines step by step each stage in the pilot's recovery process and describes the activities that must be followed if the airman expects to receive an FAA exemption for his or her condition. Program participants are required to designate a specific CRP, which consists of an itemized statement of the activities in which the person intends to engage to fill the void left when alcohol is removed from his or her life.

TWA program staffers meet with treated employees once they are ready to return to work to review the CRP and to clearly outline the company's expectations for its employees. Pilots who have undergone appropriate and successful rehabilitation may be recertified and returned to the cockpit after full disclosure to the FAA.

Cost

The costs of the program are absorbed by the company, as with other illnesses, under the Basic Group Health Care Plan, that is, 100 percent of the first $5,000 (for hospitalization), 80 percent thereafter, and paid physician visits in accordance with a Group Health Plan schedule (the patient pays the difference between the physician's actual charge and what is covered per the schedule).

Evaluation

Although there have been no published reports to date, TWA is currently conducting a study limited to the pilot population that looks at trends in absenteeism, referral services, and medical data.

Special Features

- Strong emphasis on preventing rather than simply treating substance abuse problems.

- Insurance coverage for a wide variety of behavior-medical disorders.
- Strong management-union ties in regard to program administration and facility monitoring.
- Long-term care component embodied in the Continuing Recovery Plan.

18

United Healthcare Corporation

Sponsorship

Management sponsors the program.

Objectives

The SAFECO Health Action Plan for Everyone (SHAPE) is a self-help program designed to assist employees on two counts: to take an inventory of their regular health-related behaviors and to change the ones that might have detrimental long-term effects. Like many similar ventures, its design reflects the notion that the individual must accept responsibility for personal health and fitness. SHAPE was developed with four overall requirements in mind: comprehensive scope, widespread appeal among employees, voluntary participation, and focus on those unfit individuals most in need of assistance.

Target Population

The program was introduced in May 1978 as a health promotion initiative for the SAFECO Insurance Corporation's 10,000 employees nationwide.

Since November 1979, the company's Health Maintenance Organization—United Healthcare Corporation—has marketed SHAPE to other interested corporations.

Program Structure

SHAPE is essentially a self-help venture that depends on employee interest for its success. As a result, the required staff consists simply of a program coordinator and clerical support. In addition, however, the program benefits from certain in-house resources of SAFECO Insurance Corporation, such as marketing consultation, design expertise, printing, publication of manuals, accounting/legal work, administrative support, and the assistance of two medical directors. The program has three basic components: a self-assessment questionnaire and a health-quality profile developed by health care professionals; a "how-to" notebook on life-style change; and a monthly newsletter that offers tips and encouragement on topics such as diet, smoking cessation, exercise, alcohol abuse, and stress.

Program Processes

Every six months each employee has an opportunity to answer a series of questions concerning diet, smoking, weight, exercise habits, and resting pulse rate. Participants receive a confidential health-quality profile that scores them on resting pulse rate, exercise, weight, smoking, and percentage of dietary fat. This report allows employees to compare their status to norms and to monitor their improvement over time.

Participants are also given a loose-leaf notebook containing information on behavior modification techniques, in regard to, for example, weight loss, smoking reduction, and stress management. The notebook includes a kit that helps the employee plan and implement strategies for reaching his/her health improvement goals. In addition, SHAPE participants are provided additional material, such as Dr. Kenneth Cooper's best-selling paperback on aerobic exercise, a SHAPE calorie and fat counter, and progress calendars.

Cost

The full program ("DeluxSHAPE") is sold to outside corporations based on a charge per participant for the first year (depending on the number of participants) and for subsequent years (depending on the type of package chosen and the number of participants). In most cases, employers assume the full costs of the program. However, they may elect to share costs with participating employees (usually on a 50–50 basis).

Subscriptions for the monthly 8- to 12-page newsletter ("SHAPE Write-Up") may be purchased in lieu of the full program, based on a charge per employee per year, depending on the number ordered. Larger corporations may order customized newsletters or may purchase the negatives and print the publication themselves.

Evaluation

A recent corporate survey of SHAPE participants was made in the SAFECO home office, with 56 percent of employees returning the questionnaire. The survey revealed that since they had enrolled in SHAPE, 88 percent had either lost or maintained weight, while a similar percentage claimed to carefully monitor their fat intake; 25 percent of smokers had reduced their smoking "substantially" as a result of the program and 7 percent had quit smoking altogether; and 63 percent had either continued or adopted a regular program of exercise. One of the outside firms (General Telephone Company of the Northwest) that uses the SHAPE program reported similarly encouraging findings. Sixty percent of those participants reported that the program has spillover benefits in that it influences the life-styles of both family members and friends.

On the basis of results like these, SHAPE representatives conclude that a simple and inexpensive program such as theirs can produce dramatic results. Further, they claim that their "unscientific" program rivals the more expensive company initiatives (based on the medical model) in motivating employees to make appropriate life-style changes.

Special Features

- Program marketed to other businesses.
- Strictly an employee self-help program, but may be supplemented with other fitness/health programs if a particular company chooses.
- Material available for employees' spouses ("Spousekit").
- Nonclinical approach to employee fitness.
- Designed as a long-term, ongoing program for the company and its employees.

19

Weyerhaeuser Company

Sponsorship

Although management has sponsored the various health promotion and disease prevention activities at Weyerhaeuser since 1972, representatives of groups such as the American Cancer Society and the American Heart Association often make presentations and aid screening free of charge.

Target Population

All Weyerhaeuser employees at the three sites where programs are offered may participate, and family members are welcome to use the fitness center on evenings and weekends. Approximately 1,000 of the 3,000–4,000 eligible employees are involved with one or more of the programs.

Program Structure

Three full-time staffers with backgrounds in management of physical education recreation and in exercise physiology are involved with the fitness programs. Weyerhaeuser uses employees on a part-time consulting basis to help organize and conduct special sessions. The overall program consists of the following activities: cancer awareness classes, hypertension screening, fitness evaluation, limited risk factor analysis, substance abuse programs, weight and nutrition classes, and training in cardiopulmonary resuscitation.

Used with the permission of Weyerhaeuser Company.

Cost

The annual operating expenses, excluding fixed costs, run approximately $100,000. Company employees who are interested in joining the exercise program directed by fitness center staff pay $143 per year, which allows them access to racquetball and handball courts, exercise areas, and a fitness evaluation. These members are often able to take special classes free of charge; nonmembers must pay a nominal fee.

Evaluation

Weyerhaeuser has monitored physiological changes, especially differences in conditioning level, in exercise program participants. A long-term study that would focus on such variables as changes in absenteeism and productivity is currently being planned.

Special Features

- Strong endorsement and assistance by representatives of outside groups.
- Family members eligible to use facilities during nonpeak periods only.
- Corporate interest in long-term study to assess the return on program investment.

20

Xerox Corporation

Sponsorship

This program is sponsored and backed by management, although its design and content are heavily influenced by employee suggestions.

Objectives

The directed fitness activities at the Leesburg Xerox International Center for Training and Management Development are structured to emphasize the importance of slow and progressive physical development and to focus on cardiovascular conditioning and joint flexibility. From a short-term perspective, the physical fitness and recreation center is designed to provide company employees with a means of relieving stress and tension while attending the facility training sessions.

Target Population

All trainees and the recreation center staff, as well as their family members, are eligible to use the recreation facility and to participate in the programs being offered. As many as 1,000 employees train at the Xerox training and development complex at any given time.

Program Structure

The recreation complex offers a wide variety of indoor and outdoor activities, including an exercise room, squash and handball courts, basketball and volleyball courts, jogging track, a football/soccer field, a 25-yard swimming pool, badminton courts, tennis courts, and an 18-hole putting green. The program director is assisted by a small professional staff trained in leisure sports, recreational activity, physical fitness, and exercise physiology.

Some of the directed programs such as swimming instruction, as well as facilities such as the putting green, are available during certain seasons only, yet the vast majority of activities are available through the year. Although employees always have an abundance of programs from which to choose, the staff regularly evaluates and changes the offerings to meet different employee group interests and needs.

The Xerox Health Management Program provides a "Fitbook" to interested employees. This manual contains a four-module series of exercises designed to improve muscular strength and endurance, joint flexibility, and cardiovascular reserve. Diet and nutrition guidelines, relaxation techniques, and special back exercises are also included in the book. The hazards of smoking and substance abuse are outlined and guidelines for changing individual behavior are provided. Finally, a 28-minute wellness-oriented film entitled "Health and Lifestyle" discusses risk factors and health behavior modification techniques.

Program Processes

The trainees receive introductory material concerning the recreation program as part of an overall orientation packet before they arrive at the training facility. Once they arrive, prospective recreation program participants receive a brochure that lists available activities, describes the court reservation policy, maps out both indoor and outdoor recreational facilities, and contains a program evaluation form. Participants complete a health and fitness questionnaire as the first step after enrolling at the fitness center. An exercise specialist is then available to design and recommend a personalized cardiovascular fitness program.

Calendars and tournament schedules vary depending on the size and duration of the training sessions, as well as on the participants' interests. The recreation staff may instruct interested employees in nearly any sport for which the center maintains facilities. The top two enrollees receive special recognition certificates as each competition activity is completed.

Self-tests and questionnaires designed to help indicate a participant's health and fitness status are included in the "Fitbook." The results of these self-administered tests enable an employee to select those areas of concern that seem to deserve particular attention. The participant contracts with himself/herself to reach certain fitness and conditioning goals within a three-month period. A self-addressed pledge card is mailed to the participant at the end of that time so that he/she can monitor progress and be reminded of the ongoing need to exercise.

Cost

The annual operating costs, not including overhead, maintenance, and depreciation, run approximately $150,000 per year.

Evaluation

All fitness programs such as slimnastics, swimnastics, and dancercise have a program evaluation at the completion of the course. A three-month follow-up program evaluation is designed so management can see if participants' fitness needs are being met.

Special Features

- Short-term program with follow-up three months later.
- Program content strongly influenced by employee recommendations on a continual basis.

Of the 12 background papers presented here, all but the last one (Chapter 32) are adapted from papers prepared for the National Conference on Health Promotion Programs in Occupational Settings, held January 17–19, 1979, in Washington, D.C.

The papers offer a comprehensive survey of current approaches to health promotion programs at companies all over the nation. Six papers (Chapters 25–30) are devoted to specific components of the programs. The other 6 papers discuss the problems involved in promoting health behavior change not only in general but in relation to occupational hazards (Chapters 24 and 31, respectively); the techniques of health risk appraisal (Chapter 22); the pros and cons of health intervention programs in industry (Chapters 21 and 23); and the complexities of evaluating cost-effectiveness (Chapter 32). The papers are replete with cautionary advice, encouraging data, and helpful guidelines.

All the conference papers were supported by contract No. 282–78–0174 with the Office of Health Information, Health Promotion and Physical Fitness and Sports Medicine, Public Health Service.

III

BACKGROUND PAPERS

21

Perspectives of Industry Regarding Health Promotion

G. H. Collings, Jr., M.D.

It is becoming increasingly clear that the most attractive opportunities for real health improvement in this country do not lie in the governmental arena. Disenchantment with Medicare, Medicaid, and other recent programs, coupled with a changing political climate, would indicate that the solution to our national health care problems will probably not be forthcoming from one grand health superstructure at the federal level. It is also unlikely that organized medicine will rise to the occasion and provide the working basis for comprehensive health improvement. Our community-physician-hospital illness care system will unquestionably continue to improve the currently excellent care it provides for episodic illness; and doubtless it will add some broadening of focus to incorporate more health promotion functions. But there are embedded constraints in this system that limit its universal applicability and will seriously blunt its contribution to real health improvement at least for the next five to ten years.

This leaves *business* as the principal remaining structure in which the desired health objectives can be reached. It is only recently that the opportunities for health improvement that exist in the workplace have begun to be recognized, but happily there is now a rapidly growing awareness that this is where many favorable factors converge to produce a climate highly suitable for fruitful exploration.

Dr. Collings is Corporate Medical Director at New York Telephone.

Business itself has only lately become conscious that it is a significant stakeholder in the health of its employees and their families. The "attention getter" was of course the escalating cost of health benefit plans, which now involve hundreds of millions of dollars annually for many large corporations and equally significant segments of corporate resources for smaller organizations. As a result of these economic pressures, business is overcoming its natural reluctance to get involved in health care matters and is finding that corporate opportunity in the health area transcends the original issue of cost containment. Among the advantages are less absenteeism and more productivity on the job; reduced coping problems among employees (both management and nonmanagement); and enhanced functional efficiency of the corporate organism as a whole. Although the methods for obtaining these ends may as yet not be fully apparent, one can hardly take a serious look at the health scene from the corporate viewpoint without intuitively sensing the huge potential that awaits development in this area.

Now what is it about the work environment that makes it so attractive for health system purposes?

1. More people (particularly women) are in the work force today. It is common for two or more members of a family to be working. As a result, reaching a majority of the entire population through the workplace is a practical possibility.

2. In spite of increased job mobility these days, the majority of the work force is very stable. This means that an employee tends to stay with one employer for a long time—perhaps even an entire working lifetime. This loyalty provides the opportunity to apply long-term interventions for health improvement.

3. The periodic acquisition of health data can be accomplished with relative ease, thus making possible:
   a) Study of the natural history of health and disease
   b) Tracking of individual health behavior against expected group norms and against program objectives
   c) Evaluation of individual programs and/or complete operating health systems

4. In general, workers are willing to participate in health programs offered at the worksite. Experience has shown that a much higher degree of voluntary participation may be expected in health service offered at the worksite than that same service would command at other locations. For example, multiphasic screening programs in industry regularly

achieve 90 to 95 percent participation, whereas identical programs offered free and after extensive publicity in a community setting will do well to get 30 percent participation.

Some of the reasons for the willingness of employees to participate are:

a) Convenience. Generally the industrial health program is located near to or at the person's job site and therefore requires a minimum of scheduling and travel to take advantage of it.
b) No cost. This is, of course, not unique to worksite programs, but there is a particular psychology widely prevalent among workers that what is received free at work is really part of their pay for working and as such is an entitlement that they do not wish to lose. Consequently, there is strong psychological pressure to utilize free services when offered.
c) The attitude of quality. Although many employees do not regard all that their company does with the highest degree of approval, there is an almost universal attitude that if the company provides something, the quality must be good or the company would not have "bought" it. Thus a substantial proportion of people's usual natural suspicion of do-good programs is eliminated when these programs are sponsored by the company.

   For example, if the company offers a dental insurance program to employees, there may be widely divergent opinions as to whether the company should be offering dental protection at all and as to whether, if offered, the program should contain more or less items of dental care in its schedule of benefits. But there is little concern for the quality and reliability of the insurer or of the specific insurance contract. It is assumed that the company will have looked into these matters to check that everything is first rate. By contrast, if the employee were purchasing his own dental insurance, he would be much concerned with whether he had inadvertently selected an unreliable insurance carrier and also would consider the possibilities of hidden phraseology in the contract that might, under certain circumstances, minimize or negate his protection altogether.
d) The reputation of the company medical department can also be a big positive factor in worker participation. Where the medical department has earned a reputation for the three C's—competence, confidentiality, and concern—there is apt to be less suspicion of new programs offered by that medical department.

e) When the company permits employee participation during working hours, employees can "get out of work" by being a participant. This obviously could become a major problem if abuse of the privilege is not controlled. Nevertheless, even when abuse is not permitted, the chance to take part during work time tends to encourage participation.

5. The Occupational Safety and Health Act and other related legislation and regulations at both the state and the federal levels are rapidly extending the realm of health effects considered to be occupational in origin and are commanding a no-nonsense approach to minimizing the adverse effects of the work environment. Coincidentally, the line between so-called occupational and nonoccupational conditions is becoming increasingly fuzzy. In fact, although much medical and legal time and effort are still devoted to arguments on this score in individual cases, there is no such thing as a truly nonoccupational disease or injury if the definition of "occupational" is "arising from or contributed to by work." When as much as 8 hours out of 24 are spent at work, there are simply no medical conditions that are not influenced to some extent by that work. The concept of occupational versus nonoccupational illness or injury is consequently becoming more a matter of degree than a question of absolute distinction.

   As a consequence of these two developments, more and more of the total health of workers is being accepted as the proper province for so-called occupational health programs.

The two principal goals of any system of comprehensive health care are (1) disease correction and (2) health promotion. Health promotion can be divided into two areas that, though conceptually distinct, grade into one another: (1) disease prevention or postponement and (2) wellness* improvement. Both areas are characterized by multiple undeveloped opportunities, but because medical professionals characteristically have a disease orientation, it should not be surprising that most of the work done so far has been in disease prevention. Only recently have a small number of scattered investigators begun to explore opportunities in defining wellness states and in searching for ways to improve wellness—or what might also be called improved coping ability.

In this regard a parallel historical lesson might be noted. The highly successful American Safety movement did not get very far as long as emphasis

---

*Where "wellness" is something more positive than the mere absence of disease.

was on injuries. It wasn't until emphasis was shifted to safety that real progress occurred. Perhaps we will also find that not much progress toward health improvement will occur until we replace our disease orientation with a positive health emphasis. Certainly one could argue that several decades of effort have produced relatively few specific examples of successful disease prevention. My personal view is that this unimpressive record need not mitigate against future additional attempts to find ways to prevent disease. It does mean, however, that we should not expect all, or even most, of the needed progress toward *health promotion* to come from disease prevention techniques now or in the future. It seems to me that there are much more fertile areas for exploration in wellness improvement and that we will find an abundance of success there.

Actually the greatest progress in the future will probably come from *integrated* systems where wellness improvement, disease prevention, and disease correction are all included and appropriately balanced for each participating individual. But since such integrated systems are only in their embryo stage today *(1)*, it is more common for current investigators to deal independently with each individual intervention. Thus we see hypertension control programs developing in isolation; we see one company active in the control of obesity among its employees; another caught up with stress reduction efforts; and yet another chasing glaucoma or arteriosclerosis. Indeed, this conference has oriented its approach along just such lines with "state of the art" papers addressing each of a number of possible health improvement *fragments*.

One of the problems with this fragmented approach, of course, and one that we have not adequately addressed is the sheer impossibility for any given individual to follow all of the current health improvement recommendations, much less all of the ones likely to be available in the future. Consider what your life would be like if every day you were to try to cram in an hour for aerobic exercise, two periods of 20 minutes each of meditation, a half hour of special exercises to prevent back trouble, 15 minutes of combined bookkeeping to calculate and track the data on intake of nutrients, two periods of 7 minutes each for dental prophylaxis, a half hour for reading the latest health educational material, time to attend a session on smoking cessation, and on and on.

Quite obviously, everyone cannot do everything recommended for better health. In fact not everyone needs to do all of these things. How, then, does one select what is best for himself or herself? I suggest that this is a question that sooner or later must be addressed, and a satisfactory solution must be found if maximum effectiveness of health promotion efforts is to be obtained as a societal objective.

But that is in the future. Today we continue to follow a piecemeal approach based solely and narrowmindedly on pursuing the possible. For example, we have demonstrated that it is possible to detect hypertension and to treat it successfully. It is possible to detect cervical cancer at an early and curable stage. It is possible to counter stress with relaxation. It is possible to alter body responses by biofeedback mechanisms. It is possible to improve the quality of life by sustained moderate exercise. And so on.

The trouble with all of this is that we have often jumped to the unwarranted conclusion that because something theoretically constructive is possible, everyone should do it. In fact we immediately launch programs to convince others to launch programs to involve everyone so that maximum benefit can result to humanity. In the business and industrial arena, programs along these lines are very likely to continue to proliferate just because the work environment makes it feasible and conducive to do so.

But overall, does maximum benefit result from this approach? As a matter of fact, is this approach efficient and cost-effective and, if not, is it sustainable on a long-term basis? I think not. Considerable experience with health-improving programs in industrial populations has led me to the conclusion that generalized programs applied indiscriminately to all people are only partially successful in achieving their goals. More importantly, they are usually not cost-effective and have to be abandoned.

Consider cervical cancer as an example. With the development of the "Pap smear," early cervical cancer detection became possible. Our official agencies and institutions soon recommended programs based on the objective of universal Pap smears for all women once a year. Business was exhorted to provide such a program for women employees in its own economic interest and in the interest of the health of the employees.

Granted this is *possible.* Is it practicable? In my company, there are approximately 40,000 women, among whom on the average there are two cases of disabling cervical cancer a year. Pap smears on all 40,000 of these women would cost around a half million dollars a year or one-quarter million dollars for each cancer case. That is not very attractive on a cost-effective basis.

The problem here, of course, is the indiscriminate application to all women when all women do not have equal risk of developing cervical cancer. As an alternative, suppose we could identify a subgroup of 8,000 or 10,000 women containing the higher-risk persons. If we then performed Pap smears on these women in preference to those of lower risk, it should be possible to bring the cost per case of disabling disease prevented into more reasonable dollar ranges.

The concept involved in this example may be referred to as *risk identification.* It operates by identifying the factors that are associated with risk and by using these factors to separate the population into high-risk (or high-need) and low-risk (or low-need) subgroups, thereby reducing overall cost and improving effectiveness of outcome.

A number of examples of developments along this line have been reported in recent years. In fact, risk identification as applied to individuals and their personal health maintenance has been developed into a relatively sophisticated process. But we have been slower to apply risk identification to program planning and system development. It is always logistically and operationally easier to design and carry out programs where a standard procedure is done in a standard way for everyone. By contrast, discriminating on the basis of risk or need is not only more complicated but requires definitive risk information that as often as not is only partially available. In spite of these difficulties, though, enough work on risk identification has been done to show that the principle is sound and moves the cost-benefit ratio in the right direction. Unfortunately, in many cases it doesn't move it far enough and many modalities that are perceived to have potential for health improvement still do not achieve effort-outcome ratios sufficiently attractive to make them competitive for available dollars. In many if not most cases, it is therefore becoming apparent that successful strategies for health improvement must go one step beyond risk identification to achieve what we are now beginning to call *yield potential.*

The yield potential concept is based on the commonly recognized fact that for any given intervention modality (e.g., the treatment of hypertension) all high-risk (hypertensive) persons do not respond with equal success to its application. Obviously, patients who are able to attain and maintain a normal blood pressure will be greater successes than those who respond poorly or not at all. If the identifying characteristics of patients who will succeed can be specified in advance, priorities in the application of the modality can be used as a further refinement in the pursuit of favorable cost-effectiveness.

Thus as an overall strategy for health promotion we arrive at a combined *high risk–high yield* philosophy. Though complex, this philosophy permits the development of systems that apply given modalities to target populations at frequencies and in ways selectively varied for individuals, depending on risk *and* yield factors that will produce a maximally acceptable and cost-effective program.

This and similar strategies are likely to characterize health improvement efforts in the near future. They will flourish in the work environment. They will be attractive to participating employees, to unions, to management, and to government alike, for they will not be applied to those who don't need them and will not be onerous to those who do. And overall they will produce benefits that are multiples of their cost. Through such mechanisms, particularly in the workplace, I think we can look forward to significant improvement in the real health of our people.

REFERENCES

1. *Health—a corporate dilemma; health care management—a corporate solution: Background papers on industry's changing role in health care delivery,* edited by R. Egdahl and D. C. Walsh. Springer Series on Industry and Health Care, no. 3. New York: Springer-Verlag, 1978, p. 17.

22

Health Risk Appraisal

Axel A. Goetz, M.D., M.P.H., Jean F. Duff, M.A., M.P.H., and James E. Bernstein, M.D.

Health risk appraisal is both a method and a tool that describes a person's chances of becoming ill or dying from specific diseases. The procedure generates a statement of probability, not a diagnosis.

RISK ESTIMATION

If one can accurately estimate a person's risk of (1) getting specific diseases, both physical and mental (e.g., having a myocardial infarction), (2) dying from certain diseases (e.g., from breast cancer), and (3) dying within a defined period (e.g., within 10 years), then it is appropriate to ask of what use these estimates are. To answer this question adequately, it is essential to recognize the underlying hypotheses that require substantiation in the use of health risk appraisal.

Hypothesis 1. Given a particular disease with a known incidence and for which there are identified risk indicators, a change in the prevalence of these risk indicators in the population will result in a change in the incidence of the disease.

There are two versions of hypothesis 1: the full-benefit assumption and the partial-benefit assumption.

This article was first published in *Public Health Reports* (March–April 1980), pages 119–126, and is abridged and otherwise slightly altered here. The authors are with General Health, Inc., Washington, D.C. Dr. Goetz is Vice President for Research and Development, Ms. Duff is Vice President for Programs, and Dr. Bernstein is President.

Full-benefit assumption: The change in disease incidence resulting from a change in the prevalence of a risk indicator will reflect the full benefit from this change (for example, a nonsmoker and a just-stopped smoker will have the same risk of myocardial infarction).

Partial-benefit assumption: Only partial benefit is derived from a change in the prevalence of a risk indicator (for example, a nonsmoker has less risk of a heart attack than a just-stopped smoker).

Hypothesis 2. Giving people information about their own risk will lead to actions perceived as, and directed at, reducing risk.

Neither hypothesis 1 nor 2 has been fully tested. Most recent investigations have focused on hypothesis 2 on the assumption that hypothesis 1 is valid. Although hypothesis 1 remains unsubstantiated, pursuit of programs designed to reduce the prevalence of risk indicators in defined populations seems prudent. Hypothesis 2 can be tested only if risk estimates can be conveyed to individuals. However, little is known about the most appropriate methods for transmitting such information—and much is known about how information alone will not result in changed behavior.

Our purpose here is not to present evidence in support of risk appraisal as a tool for behavioral change but rather to address the present state of risk estimation.

RISK ESTIMATION VERSUS DIAGNOSIS

Medicine traditionally focuses on the diagnosis and treatment of disease. The patient's medical history, physical examination, and laboratory tests are designed to provide clues, signs, and symptoms that, when considered together, will suggest a diagnosis. The intent is the immediate detection and identification of disease. Risk estimation calls for a different approach to information gathering—the collection of data that will permit outcomes to be anticipated over a much longer time frame, one measured in years, if not in decades.

Blood pressure can serve as an example of the distinction between the information traditionally gathered to assess physical status and the information required to estimate risk. In conventional teaching, a blood pressure of 120/80 mm Hg is considered normal. A blood pressure of higher than 140/90 is defined as borderline hypertension and one higher than 160/95 as hypertension. A person who has a blood pressure lower than 140/90 is considered healthy. Currently, diastolic blood pressure is regarded as more important than systolic. (In the pumping of the heart, the relaxation at dilation phase is known as diastole, the contraction phase as systole.)

From the perspective of identifying persons at risk of coronary heart disease and stroke, blood pressure value can be used to estimate the contribution to the risk of these diseases. And what is more important, at levels of blood pressure considered within the healthy range by most clinicians, risk may be substantially elevated. A person with a pressure of 140/95 may have a considerably higher risk than was thought in the past.

But blood pressure is not by itself a sufficient basis for estimating the risk of cardiovascular disease and stroke. Variables including age, cholesterol levels, cigarette smoking status, Type A personality, high-density lipoprotein levels, and energy expenditure rates through exercise also contribute to cardiovascular disease risk. The interrelationships among these variables are complex and require computation that cannot be performed in most clinical practice settings.

This complexity is one of the reasons that risk assessment and both individual and aggregate group risk profiles have, until recently, rarely been used in clinical settings. This infrequent clinical use, however, is no longer a function of technical or informational deficiencies. Risk can be estimated, and those who are involved in guiding people toward healthier lives and those who have the responsibility for maximizing the health of the employed can expand their practice patterns to include an accurate estimation of risk.

To incorporate risk estimation into clinical practice, health professionals may have to gather additional information above and beyond the routine medical history. An expanded knowledge of risk indicators and risk factors, as well as computer support facilities, is required.

GENERAL APPROACH TO RISK ASSESSMENT

Suppose we were to estimate a person's risk of death within the next 10 years. Not knowing anything about the person, our best estimate would be the average risk regardless of age, race, sex, or any other characteristics that the person might have. This estimate, although expressed in a precise figure, would leave a large amount of uncertainty. If, however, we know the person's age, this uncertainty would be greatly reduced, although the estimate still would not be very useful. The more we know about a person, the better we can estimate that person's disease and death risks—up to the point at which we have asked all the questions that we know are related to risk. To the extent that a risk appraisal instrument approaches this point, we can consider it to represent the best state of risk estimation, given current epidemiologic research evidence.

Individual risk estimates are produced by modifying the data on the average risk in a population according to whether the person has or does not have certain risk-related characteristics. Such characteristics might be, for example, a history of a certain disease like diabetes, a particular habit like frequent vigorous exercise, or a test value like the blood pressure level. Any characteristic with consequences for a health-related risk is called a risk indicator.

To estimate risk, we must know not only whether a person has a set of risk indicators but also whether any given indicator is associated with an increased or decreased risk relative to the average and, further, what the magnitude of the person's deviation is from the average. This information is contained in the risk factor. A risk factor is the quantitative weight attached to a risk indicator to describe, for a particular cause of disease or death, the amount that the indicator increases or decreases the risk of death or disease. It should be noted that this terminology is but one among several. So far, no commonly accepted standard has emerged. This fact is itself a comment on the current state of risk-related research and its application. When we multiply the appropriate age-sex-race-specific average morbidity or mortality data by the combined risk factors for a person, we obtain that person's estimated risk of disease or death.

WHAT RISK ESTIMATION IS AND IS NOT

Implied in the risk estimation method just described is a comparison of the person whose risk we are estimating with groups of persons who, in the past, have shared the same risk indicator or set of risk indicators. Also implied is the assumption that the presence of the same risk indicators has the same health consequences now as it did then. It is difficult to assess the validity of these assumptions. We are therefore limited to *predictions* of risk in a statistical sense. But risk estimation is not the prediction of a person's future medical history. Even if two persons had an identical set of risk indicators, their fate might be vastly different because of variables that have not as yet been captured in the risk indicator set, such as environmental exposures presently not known to have consequences for risk, pathogenic processes under way, or differences in genotypes. Although, with increasing knowledge about risk indicators and risk factors, the range of possible health outcomes for a person can be assessed with ever greater precision, estimates of risk give only the odds or likelihood of an event, such as a myocardial infarction, occurring in a group of people with or without a certain set of characteristics (such as smoking, average weight, or low blood pressure).

If we are to make use of the growing knowledge reflected in the epidemiologic and biomedical literature and to obtain more precision in risk estimation, the risk indicators and risk factors used in our risk-assessment instruments must continually be updated. To maintain the precision achieved, the data bases for morbidity and mortality incidence must also be brought up to date regularly.

THE USE OF HEALTH RISK ESTIMATION TOOLS

Part of the process of selecting a Health Risk Estimation tool for a particular group or situation will be to consider what the instrument can contribute to an overall program. In addition to estimating risk with accuracy, and feeding information back to the program participant in an intelligible manner, a good Health Risk Estimation tool should make the following contributions to a health promotion program.

1. *Provide participants with a comprehensive overview of their health risks.* A good Health Risk Estimation tool should be capable of serving as the first step in a health promotion program. It should provide participants with an accurate estimate of mortality risk for the leading causes of death, morbidity risk for cancer and cardiovascular diseases, and at least a normative study of aspects of mental well-being. If participants can use and interpret the tool without a lot of help from staff, and if it acts as permanent record of their risk status at the start of the program, its inclusion in the program will be desirable.

2. *Identify the participants most likely to reduce risk.* The tool should include a method of assessing aspects of a person's health beliefs and attitudes so that those persons who are most likely to take steps to reduce their risks and maintain their health can be easily identified. This permits risk reduction programs to be targeted to those most inclined to use them, as well as alerting program staff to those who are both at high risk and judged to be least likely to act to care for their health. This feature of a Health Risk Estimation tool is especially useful in a controlled environment where the confidentiality of the data is protected, because it requires that group members be identified and followed up individually.

3. *Provide aggregate analysis of the data.* Users of a Health Risk Estimation tool should be able to obtain an aggregate analysis of the group's data. This analysis should cover all the basic data elements and should be capable of being tailored to the particular needs of the group. Of course, the quality of the analysis is directly determined by the quality of the data collection tool—that is, the Health Risk Estimation tool. An aggregate analysis protects the confidentiality of individual results because no identifying data are included. It can be used as a support for health program planning, as a guide for targeting health promotion dollars within a group, and as a method of monitoring changes in risk and risk-related behavior over time.

OUTLOOK FOR RISK ESTIMATION

The rapid progress of epidemiologic and biomedical research will no doubt help increase the scope and validity of risk estimation techniques. Benefits from research will be maximized if the information needed to improve risk estimation is clearly delineated.

By becoming research tools themselves, health risk appraisal instruments can contribute greatly to the knowledge base for risk estimation. Responses to appraisal instruments can be aggregated to produce large data matrices with several hundreds of variables for many thousands of persons. Such aggregation will permit the definition of a wide variety of subpopulations and the identification and study of risk-related variables. Prospective study of the persons in such a data base, combined with the analysis of data on their morbidity and mortality and health service utilization over time, would significantly increase the contribution of risk-appraisal data bases to the art of risk estimation.

Whether identification of the population and individuals at risk, coupled with programs that are known to decrease the prevalence of risk indicators, will in fact yield the dividends of healthier, more productive people and reduce expenditures related to illness, disability, and premature death is another, often controversial issue. As more risk estimation and risk reduction are carried out, we must make sure that data to answer these questions are collected and analyzed.

23

A Perspective on Health Intervention Programs in Industry

M. Donald Whorton, M.D., M.P.H., and Morris E. Davis, J.D., M.P.H.

This paper, a perspective on health intervention programs (wellness programs) and their relationship to the industrial environment, reflects solely the opinions of the authors, which are not to be construed as representing policy, understanding, or agreement with any institution or organization.

OVERVIEW AND INITIAL ASSESSMENT

The most obvious question is, Why use the workplace? There are at least four issues that are involved and that should be addressed.

- Do the employees represent a captured group of participants?
- Is there subtle coercion that insures full participation and/or cooperation?
- Is this a new frontier in preventive health?
- Are preventive health programs only successful in those cases in which there is group participation?

Dr. Whorton is Senior Health Associate at Environmental Health Associates, Inc. He is also Medical Director, and Mr. Morris is Director of the Labor Occupational Health Program, Institute of Industrial Relations, University of California, Berkeley.

In addition to these four issues, other concerns need to be addressed.

- Is there concern for health problems that affect the individual's ability to work?
- Is there concern for changing the individual's life that may also include family members?
- Is there concern to provide "comprehensive health benefits and services" with the objective of reducing utilization of health plan benefits (emphasis on preventive health)?
- Is there concern for identification and treatment of disease and illness?
- If treatment is not the objective, who will pay for the necessary treatment? (This also raises the question of various methods now used for payment for such treatment, i.e., health insurance, workers' compensation insurance, union negotiated differentials, social security disability, unemployment insurance.)
- Is rehabilitation an objective, depending on findings?
- Are there restrictions for participation in such intervention programs? For instance, is eligibility based on age, duration of employment, seniority, sex, or on abstinence from smoking, drinking, or, drugs?
- Have the objectives, problems, and ultimate intentions of the intervention program been adequately and reasonably considered before implementation in order to avert questionable results?

POTENTIAL PROBLEMS IN USING THE WORKPLACE FOR INTERVENTION PROGRAMS

Traditional View of Occupational Medicine

Traditionally in the United States the health care provided at the workplace has been limited to occupationally related illnesses or injuries. Nonoccupational illnesses or injuries are reserved for treatment by the individual's personal physician except during emergencies or for "courtesy treatment." The late Dr. Henry Howe described the objectives of an occupational medicine department as follows:

> The basic objectives of an occupational health program are succinctly stated in the 1971 revision of the AMA's "Scope" as follows: "To protect employees against health and safety hazards in their work situation.

> (2) Insofar as practical and feasible, to protect the general environment of the community. (3) To facilitate the placement of workers according to their physical, mental, and emotional capacities in work which they can perform with an acceptable degree of efficiency and without endangering their own health and safety or that of others. (4) To assure adequate medical care and rehabilitation of the occupationally ill and injured; and (5) to encourage and assist in measures for personal health maintenance, including the acquisition of a personal physician whenever possible." *(1)*

Dr. Howe described the relationship between the occupational physician and the community physician in this way:

> The plant physician should establish and maintain the best possible relationships with the family physician of the plant's employees. The plant physician, when evidence of nonoccupational illness is discovered upon the employee's physical examination or dispensary visit, should, with the employee's consent, refer such evidence promptly to the employee's family physician. The plant physician should never use his industrial affiliation improperly as a means of gaining or enlarging his private practice. *(1)*

It is highly probable that a sizable percentage of occupational physicians would view any type of health intervention program not specifically related to an occupational etiology as outside the scope of their mandate. In addition, such programs could be vigorously resisted by an employee's personal physician or by local medical groups.

Whose Agent Is the Occupational Physician?

Who will conduct and/or control the intervention program? Will it be the employer's medical department or an outside organization? If it is done by an outside organization, then as whose agent does that organization function? This is a more thorny question than might at first be suspected, as is evidenced by the debate and discussion over the issue of whose agent the occupational physician is.

> In occupational medicine there is a real or perceived problem about loyalty of the physician. A review of the literature reveals a variety of views on where the loyalty of the occupational physician rests. Many of the articles raise a question of primary loyalty—is it to the individual, the corporation, the government, or an academic entity?

Dr. Irving Tabershaw, in an article entitled, "Whose 'Agent' Is the Occupational Physician?" *(2)*, points out that:

1. "[The physician is] a professional . . . he works for no other purpose than the benefit of his patients. In occupational medicine that is the worker."
2. As with other third-party payers, the patient is still primary.
3. The physician is *not* an "agent of industry."
4. The medical records are the patient's via his/her physician.

In another article, "How Is the Acceptability of Risks to the Health of the Workers to Be Determined?" *(3)*, Dr. Tabershaw states that:

1. The physician must inform the workers about the hazards of the job in language they can understand.
2. The physician is responsible for the individual worker.

Dr. Merle Bundy, in "How Do We Assure That the Workers' Health Is the Occupational Physician's Primary Concern?" *(4)*, makes these points:

1. The concern for the health of the workers must be shared between management and the occupational physician.
2. The physician should be neutral in labor and management matters.
3. Where reasonable doubt exists, medical issues should be resolved in favor of the employee and the maintenance of his health.

Dr. William Morton also addressed this issue in an article, "The Responsibility to Report Occupational Health Risks" *(5)*. He suggests that:

1. Some occupational physicians are restricted from communication about job hazards by their employers by such factors as fears of liability and increased capital costs, among others.
2. Some physicians feel their primary allegiance is to the provider of the paycheck—be it private, governmental, or educational.
3. Company physicians have a bad name, even worse than that of organized medicine. *(5)*

Workers' View of Occupational Physicians

In addition to the professional grappling with whose agent the occupational physician is, many workers wonder whose agent he is.

How do workers view the occupational physician? Although the following information is not the result of a scientifically and statistically valid study, it constitutes a summary of general impressions gleaned from talking with various workers—from the individual rank and file to union officials at various levels. Workers expressed feelings about the "company doctor" which ran the gamut:

1. The physician is concerned about their health and thus the workers are not afraid to tell him anything.
2. The physician is a "good guy" but his hands are tied because he must dance the company's tune.
3. The good physicians don't last long because they are forced out.
4. There is a feeling of distrust because the workers believe that the physician's primary obligation is the welfare of the company and not that of the workers. The workers never want to be evaluated by the physician because of the perceived belief that adverse medical information will be used against them in a discriminatory manner.
5. The physician is called "bum," "quack," "veterinarian," or some other disparaging term.

> There appears to be a much more negative feeling toward the occupational physician than toward the occupational nurse, industrial hygienist, or safety engineer. This probably is a reflection of what patients (workers) feel about their personal physician. Although the public at large is perceived generally to have negative feelings toward organized medicine, individually most Americans have trust and confidence in their personal doctor. It is apparent that trust in the personal physician does not always extend to the occupational physician. (6)

Other Issues

Of extreme importance is which parties will receive the results of an intervention program and how this information will be communicated. Before the beginning of the program, it must be made clear how each of the following parties is to be informed as well as the type of information they are to receive:

1. The employee
2. The employer
3. The employee's personal physician
4. The employer's physician
5. When appropriate, the employee's collective bargaining agent

In any workplace program, the stigma of refusal to participate can have serious ramifications. There can easily be subtle as well as overt pressure, such as potential job advancement, to assure a high percentage of cooperation and participation. Obviously, certain types of programs, such as substance abuse programs, may be the employee's last chance to save his or her job. The question of refusal in that situation would be viewed differently than in the case of a person whose job was not in jeopardy but who did not want to participate in the program for personal reasons.

An employee might refuse to participate in any type of voluntary health program associated with the workplace because he fears that the employer could discover personal information about him. In essence, the employee would view this as a real or perceived loss of privacy.

In addition to the refusal problem, one must address the possible effects of an adverse finding or outcome from the program. If the employee's job risks being placed in jeopardy by findings or outcomes that are not favorable, then job security is certainly a key to employee participation. A recent example of concern about the job security aspect in participation in periodic medical evaluations is reflected in the OSHA lead standard. Certainly, this standard does not resolve all of the problems, as can be witnessed by subsequent legal actions by involved parties. On the other hand, commercial airline pilots must undergo and pass mandatory periodic health examinations in order to fly. In this situation concern for the public's health outweighs the individual's right to privacy.

Another concern is that the intervention program may take the place of true occupational health or environmental control programs. Care must be taken to assure that the intervention program does not stifle or inhibit occupational health programs that are concerned with the health effects of work. An example of such a situation is the previous belief of the State of Maryland Health Department that collecting blood for serology or conducting tine (tuberculosis) tests at workplaces (because of captured populations) was the adequate extent of the state's occupational medical program *(7)*.

MEANING AND USEFULNESS OF INTERVENTION PROGRAMS

Three aspects of medical screening programs or medical procedures will be briefly discussed as prototypes of intervention programs. The general points may not apply to all intervention programs.

Overview

In medical screening programs of procedures, there is the obvious difference between the symptomatic patient and the asymptomatic patient. The effectiveness of most tests and procedures has been validated on symptomatic or sick individuals. Frequently outcome studies of the same tests or procedures on asymptomatic individuals either have not been done or are inconclusive.

> Spitzer and Brown *(8)*, in an article "Unanswered Questions about the Periodic Health Examination," focused on three areas: the impact on health; the content of a beneficial health examination; and the effect of the examination on the patient/physician relationship. In the past, the establishment of a diagnosis was the main objective or the desired outcome of a "history and physical" or a periodic examination. The current professed emphasis is an outcome which identifies those disease states for which something can be done to improve health, improve quality of life, extend life, or even "cure" the problem. Little emphasis is placed on finding disease for which nothing can be done to change its natural history.
>
> Is screening beneficial to the patient? There is not conclusive evidence that identification or intervention alters the course for most adults. Data from the 50's and early 60's showed positive results from periodic examinations; however, the data were biased due to the socio-economic groups that were examined. Data from the current decade are also not conclusive. Mortality data are suggestive but not significant, while some morbidity data are reversed from what one would expect: those screened have more problems than those not screened. *(9)*

Hypertension Screening Programs

Alderman and colleagues have reported successful results in the detection and treatment of hypertension at the worksite.

> A program linking detection to treatment was designed to improve blood pressure control among adults with asymptomatic, uncomplicated hypertension. Key elements of this program were provision of all diagnostic and therapeutic services at the work site, integration of delivery system with administration of the labor union, adherence to a rigid protocol, and continuous patient surveillance by nurses and paraprofessionals. *(10)*

As for the cost-effectiveness of such a program, the authors note:

> To balance this investment industry must anticipate that economic benefit will result from reduction of absenteeism and hospitalization. While it is too early to conclusively demonstrate in this group that the pressure decline achieved is associated with reduced morbidity and mortality, experience elsewhere indicates this to be a reasonable expectation.

> Nonetheless, among Storeworkers [members of the United Storeworkers Union], a decline in sick days and hospitalization has apparently occurred and has yielded savings that offset cost of treatment so that the net burden of the program is less than $2.00 per employee per year. This has happened in a setting where almost $400 per annum is currently being invested in health and security benefits for each employee. Further savings will accrue as major cardiovascular sequelae of high blood pressure are prevented. *(11)*

In contrast, a study reported by Haynes et al. has shown the opposite result:

> A study of hypertension in an industrial setting allowed us to confirm and explore an earlier retrospective finding that the labeling of patients as hypertensive resulted in increased absenteeism from work. After screening and referral, we found that absenteeism rose (mean $\pm$ 1 S.E.) 5.2 $\pm$ 2.3 days per year ($P < 0.025$); this 80 per cent increase greatly exceeded the 9 per cent rise in absenteeism in the general population during this period. The main factors associated with increased absenteeism were becoming aware of the condition ($P < 0.01$) and low compliance with treatment ($P < 0.001$). Subsequent absenteeism among patients unaware of their hypertension before screening was not related to the degree of hypertension, whether the worker was started on therapy, the degree of blood-pressure control achieved or exposure of attempts to promote compliance. These results have major implications for hypertension screening programs, especially since absenteeism rose among those previously unaware of their condition, regardless of whether antihypertensive therapy was begun. *(12)*

Effectiveness of Test Utilized

What are some of the problems associated with the effectiveness of testing? The examples used here involve coronary artery disease and colon and rectal cancer.

> Schor et al. looked at the cause of death in 192 executives, ages 40–59, who had previously had a comprehensive regular medical check-up within one year of death. The condition leading to death was not found during that examination in 59% of the cases. *(13)*

We do not have to remind most readers about the insensitivity of the resting electrocardiogram in detecting coronary artery disease in many persons or in predicting the onset of myocardial infarction.

Annual proctosigmoidoscopy has been touted as necessary in adults over the age of 40. Day et al. *(14)*, in 1953 reported the results of routine proctosigmoidoscopies on 2,111 adults. They found neoplastic lesions in 5% of the males and 3.7% of the females. In addition, they found another 2.7% had neoplastic lesions on re-examination a year later. These data have lead many physicians to strongly recommend a yearly proctosigmoidoscopy. Some companies and governmental agencies have gone one step further and require this to be done for certain personnel.

However, Moertel *(15)* evaluated the results of proctosigmodoscopy on 42,207 Mayo Clinic patients free from symptoms of the lower GI tract. Only 55 cancers were found in these patients (a rate of 0.13%).

The conclusion is that the test is useful in patients in which some symptoms are present. Stools for occult blood are easier to obtain and evaluate as a screening test. *(9)*

In addition the patient will return for future examinations.

These examples show that there are many potential pitfalls with screening programs. These inherent problems have prompted the World Health Organization to develop criteria for screening programs *(8)*. Although not every point will apply to every intervention program, these criteria should be considered before implementation of intervention programs.

- The therapy for the condition must favourably alter its natural history, not simply by advancing the point in time at which diagnosis occurs, but by improving survival, function, or both.
- Available health services must ensure diagnostic confirmation and provide long-term care.
- Compliance among asymptomatic patients in whom an early diagnosis has been achieved must be at a level to be effective in altering the natural history of the disease.
- The long-term beneficial effects must outweigh the long-term detrimental effects.
- The effectiveness of potential components of multiphasic screening should be shown individually prior to their combination.
- If the benefits of screening accrue to the community at large, the community benefit must withstand scientific scrutiny.
- The cost-benefit and cost-effectiveness characteristics of mass screening and long-term therapy must be shown.
- The burden of disability for the condition in question must warrant action.
- The cost, sensitivity, specificity, and acceptability of the screening test must be known. *(8)*

RECOMMENDATIONS FOR SUCCESS OF INTERVENTION PROGRAMS

The following twelve points are the authors' recommendations. They are not all-inclusive but are considered to be the more cogent ones.

1. Top management must support the program with the following:
   a) An explicit policy statement for both the initial and the continuing commitment to the program
   b) Adequate staff and facilities
   c) Adequate "insurance" protection for participants
2. If a union represents the employees, it should support the program.
3. There must be incentives for participation in the program (e.g., adequate time off, loss-time wages).
4. There must be a clear understanding as to who has financial responsibility for abnormal outcomes that are beyond the scope of the program.
5. The intervention program should, when appropriate, meet the WHO criteria for periodic screening. If the program is experimental, it should be labeled as such.
6. There must be a general educational program or focus before the beginning of the intervention program and during the implementation stages. The union or workers' representatives should be involved in planning and conducting educational programs.
7. Employees should be able to participate freely without coercion. However, in situations in which substance abuse programs exist, the employee may be required to participate in the program if the substance abuse problem has placed his/her job in jeopardy.
8. The employee must be informed of the program's protocol as well as of his personal results in an understandable manner or language.
9. The employee's treating physician should receive a copy of the program's results in order to allow him full knowledge about the patient.
10. The general practice of confidentiality must be assured and observed.

REFERENCES

1. Howe, H. F. Organization and operation of an occupational health program. *Journal of Occupational Medicine* 17:360–400 (1975).
2. Tabershaw, I. Whose "agent" is the occupational physician? *Archives of Environmental Health* 30:412–416 (1975).
3. Tabershaw, I. How is the acceptability of risks to the health of the workers to be determined? *Journal of Occupational Medicine* 18:674–676 (1976).
4. Bundy, M. How do we assure that the workers' health is the occupational physician's primary concern? *Journal of Occupational Medicine* 18:671–673 (1976).
5. Morton, W. E. The responsibility to report occupational health risks. *Journal of Occupational Medicine* 19:258–260 (1977).
6. Whorton, M. D., and Davis, M. E. Ethical conduct and the occupational physician. *Bulletin of the New York Academy of Medicine* 54:733–741 (1978).
7. Whorton, M. D; Pierce, J. O.; and Radford, E. P. A program for control of occupational health hazards in Maryland. Report for the Commissioner of Labor and Industry, Johns Hopkins University, 1973.
8. Spitzer, W. O., and Brown, B. P. Unanswered questions about the periodic health examination. *Annals of Internal Medicine* 83:257–263 (1975).
9. Whorton, M. D. Periodic medical examinations: A response. Paper presented at the Appalachian Laboratory for Occupational Safety and Health Labor-Management Seminar on Occupational Respiratory Disease, June 30, 1977, Morgantown, West Virginia.
10. Alderman, M. H., and Schoebaum, E. E. Detection and treatment of hypertension at the work site. *New England Journal of Medicine* 293:65–68 (1975).
11. Alderman, M. H., and Davis, T. K. Hypertension control at the work site. *Journal of Occupational Medicine* 18:793–798 (1976).
12. Haynes, R. B., et al. Increased absenteeism from work after detection and labeling of hypertensive patients. *New England Journal of Medicine* 299:741–744 (1978).
13. Schor, S. S.; Clark, T. W.; and Parkhurst, L. W. An evaluation of the periodic health examination. *Annals of Internal Medicine* 61:999–1005 (1964).
14. Day, E.; Rigney, T. G.; and Beck, D. F. Cancer detection—an analysis and evaluation of 2,111 examinations. *American Journal of Hygiene* 57:344–365 (1953).
15. Moertel, G.; Hill, J. R.; and Dockerty, M. B. The routine proctoscope examination: A second look. *Mayo Clinic Proceedings* 41:368–374 (1966).

24

Health Behavior Change at the Worksite: A Problem-Oriented Analysis

Joseph H. Chadwick, Ph.D.

INTRODUCTION

Scope and Objectives

The purpose of this paper is to identify and discuss critical issues that cut across the specific functional elements (e.g., smoking cessation, blood pressure control, stress management) of health promotion programs at the worksite. What is wanted here is equivalent to what I would call a preliminary statement of the problem from a *systems point of view.* It is this point of view that is missing after all the component topics have been addressed. Some of the generic questions that need to be considered in such a problem statement are as follows:

1. What are the essential elements of a health behavior change system at the worksite? What problems are apparent in these elements?

Dr. Chadwick is with American Healthway Services, Menlo Park, California. At the time this paper was prepared, he was Director of the Health Systems Program, SRI International, Menlo Park.

2. How do these elements go together? What are some alternative arrangements? What are the effects of scale?
3. In the system as a whole, where are the problems, gaps, and barriers, and what can be done, respectively, to solve, bridge, and surmount them?
4. What are the common requirements, operations, and problems shared by the specific functional elements?
5. What are the critical issues in coordination? Where are the critical points of interaction?
6. What are the utilitarian objectives of the system? How can they be stated in ways that can provide guides to action? Where are the pitfalls in utility evaluation?
7. What are the costs and benefits? To whom do they accrue? How can they be measured or estimated in practice? How can they be optimized?

As will be seen in the subsequent discussion, these generic issues can easily be translated into the specific issues that relate to health promotion per se.

The paper is *prescriptive* (i.e., concerned with design) rather than descriptive in nature. It is oriented to the future and not to the past (so almost no review of the background of the subject is provided).

Since this paper is written from a *problem-oriented* point of view, it may seem to be somewhat pessimistic in tone, but that is not the intent. I believe that we are at the very beginning of this subject, and that there is a great deal to be done to convert the promise into reality. At this time, I believe that the focus should be on problems rather than promises. But I am basically very optimistic that progress will be swift if we adopt a problem-oriented approach and make a substantive commitment to problem solving. If we assume that we already have the answers, and leave things to grow like Topsy, as is usually the case with innovations in the U.S. health system, then I would have to be rather pessimistic regarding the outcome.

HEALTH BEHAVIOR CHANGE AT THE WORKSITE: CONCEPT AND REALITY

Ideal Environment

In the not-too-distant past most health care, such as it was, was provided in the home. Over the last few decades, the tendency has been to move further and further away from this convenient local service toward highly centralized service at a doctor's office, medical clinic, or hospital. This is advantageous for certain types of problems but is highly disadvantageous for an important part of the services that are needed in the prevention of chronic disease. In fact, if one is limited to these conventional modern settings, it is difficult, if not impossible, to define a cost-effective approach to the prevention of several of the most important chronic disease problems. This creates a dilemma just at the time when we are beginning to put major emphasis on such problems.

What features in settings are needed that would be appropriate for an attack on these chronic disease problems among adults? The following seem to be particularly important.

1. The setting should be physically very near to patients, because the costs of transportation and lost time are far from negligible in the situations with which we are concerned.
2. It should be physically near to everyone, insofar as possible, because even the costs of time lost in screening and rescreening are not negligible.
3. It should be appropriate for providing medical care of a nonacute, ambulatory nature.
4. It should be convenient for health surveillance and follow-up in the wake of initial entry into treatment.
5. It should facilitate mass communication and individual information-seeking, since health education is an important role in this general area of health.
6. It should be one in which it is possible to provide social support, because behavioral factors tend to play an important role in this general area of health.

There is probably no setting that fulfills the above requirements better than an industrial setting. This, of course, will not provide access to all the adult population. It could provide the basis for access to approximately one-third of the adult population, the one-third that goes to a fixed place of work every weekday.

The industrial setting also has certain very important economic and administrative advantages. To a considerable extent, the worksite is a cost and benefit aggregation point. A sizable portion of the costs of health care and of disability and death are paid at the worksite, and hence a sizable portion of the benefits from effective disease control tend to return there. According to recent calculations, the employer's costs for life insurance and health insurance plans increased eightfold in the last two decades, when *computed in constant dollars*. This great increase has occurred because both insurance coverage and the employer's share of this coverage have risen sharply over this time period (*1*).

With employers having an economic stake of this magnitude, there is a very strong business logic to health promotion programs. Analysis indicates that some program elements, carried out in an efficient manner, can more than break even on a cash-flow balance sheet. That is to say within a few years the cash benefits of these programs will equal or exceed the cash costs, entirely disregarding the intangible benefits, which are not insubstantial.

Besides providing an economic focal point in relation to health problems, the worksite also provides a very convenient management focal point. Both management and, if available, the unions tend to have highly developed management structures in being, and highly developed relationships with the employees. These structures and relationships in being can greatly facilitate the delivery of the kind of care that is needed for the control of several important chronic disease problems.

From these several points of view the worksite is so ideal as an environment for certain aspects of chronic disease control that the implications are almost revolutionary. Certainly, an enlightened use of this environment could be a major factor leading to progress in the amelioration of cardiovascular disease. This is the potential. In practice, we still have a very long way to go.

Undeveloped Potential

The potential of this approach to health care stems from the existing organization of the work force. In broad outline, what are we looking at? There are approximately 100 million persons in the total labor force, representing two-thirds of the noninstitutionalized adults. About 75 million of these persons are employed by the 2 million corporations active in the United States. Hence, the overall average number of employees per corporation is 30 to 40.

In the first 500 manufacturing companies the number of employees averages about 30,000 employees per company, and in the second 500 manufacturing companies, about 4,000 employees (*2*). Obviously, not all of the employees of most large companies are to be found at a single site. From figures of this type one can deduce that not more than one-quarter of the labor force employed by corporations is in groupings of more than 100 employees per site. At least three-quarters is in groupings smaller than 100 employees.

Most of the publicized activity in health behavior change at the worksite has been taking place among the larger companies, especially the largest 500 manufacturing companies. Even among this group of companies the percentage having such programs appears to be quite low. Exact figures are hard to come by, but, leaving out executive physicals, I would estimate that not more than 10 to 15 percent of the first 500 companies have such programs. Among small companies the percentages are much less. It therefore appears that the percentage of U.S. employees now covered by health promotion programs, excluding executive physicals, is certainly less than 5 percent, and might be no greater than 2.5 percent.

Even where such programs exist, they usually do not include all the ingredients needed to produce cost-effective results and to justify the efforts involved. Many have been undertaken on the enthusiasm of the moment, without a carefully defined, quantitative basis of action. Very few of these programs have been evaluated. In most cases, a basis for evaluation cannot be defined. This situation reflects an almost total lack of detailed information on ways and means to carry out effective programs.

All in all, the potential of the worksite for health behavior change programs is substantially undeveloped at present.

The Reasons Why

There are several reasons for the very large difference between concept and reality in this area. To begin with, health promotion programs at the worksite tend to deal largely with the problems of chronic disease control. But we are not very clear on how to deal with these problems effectively in conventional settings. In their early stages chronic disease problems are frequently asymptomatic, and almost always low-grade and prosaic in appearance. With our emphasis on the immediate and the spectacular we tend to underestimate their importance.

For the same reasons, we tend to underestimate the complexity of these problems. We are too satisfied with superficial approaches and half-solutions. But these approaches will not produce the cost-effective results that we seek. Instead of many superficial analyses, what we really need are a few much deeper analyses of the problems—analyses that many can use. It appears that good solutions should generalize to large segments of the employee population.

The situation is somewhat analogous to industrial production itself. The development of a single modern automobile, along with its associated production tooling, is a tremendously complex and expensive endeavor. Once accomplished, this development provides a basis for large-scale and highly economical production of the desired goods. Because the control of chronic disease has some common features with modern mass-production viewpoints, the worksite has one more point of logic in its favor as an environment for health promotion programs.

There is, however, another side to this issue, and its implications at first glance are negative. Health care is in fact a business, but it is not the business of most of the companies with which we are concerned. Health care programs will be the most effective when they come under the direction of senior, creative management, are capitalized appropriately, and are in all ways treated like a first-class business. But there is a limit to the amount of such resources that a company can, or should, devote to an area that is basically a side issue to their main line of business.

The service base required to amortize the development of a comprehensive set of health promotion programs and needed auxiliary apparatus has been estimated as not less than 100,000 employees (3). This puts the development of a comprehensive program beyond the reach of most large companies and all small companies, if they undertake to do this for themselves. What seems to be the most practical is the evolution of service companies, preferably fairly large companies, devoted to health behavior change at the worksite as a business.

Another problem faced by the small company is that even when it finds someone to help or advise it about health promotion programs and puts an excellent program into effect, it may well not be able to collect the resulting benefits. The greatest portion of these benefits occurs as savings in the life and disability plans, and a small company does not have premiums based on experience for such plans. This tends to negate the whole concept for about 90 percent of the target population. Something will have to be done about this.

In summary, most companies cannot afford the cost to develop detailed and effective programs for themselves. Small companies cannot even provide their own services. Hence *detailed*, extensive guidelines must be generated by some source. Legislation may be needed to enable the return of benefits to small companies. Service agencies are needed to play a variety of roles in this arena. They should be large enough to stay with and back up their products. One key role for the Department of Health and Human Services (DHHS) is to provide the evaluation tools that will be needed to police the service agencies and maintain quality control.

PROBLEMS AND OPPORTUNITIES IN HEALTH BEHAVIOR CHANGE AT THE WORKSITE

What are the essential elements of health behavior change systems at the worksite? What problems are apparent in these elements? How may these problems be addressed and solved?

There are many ways to break out the elements of such systems. The breakout of elements to be used herein is shown in Figure 24.1. This particular categorization provides elements that *cut across* the specific functional threads (smoking, blood pressure, etc.) of the system. It is a categorization that attempts to be both comprehensive and balanced. Each element represents a set of problems, issues, and opportunities that are coherent within themselves, and in some way very important in relation to health promotion at the worksite. In the following paragraphs these elements are discussed, one by one.

Policy and Ethics

This is the first element in any health behavior change system, and I want to touch on a few key issues. To begin with, there is a very important paradox in this element.

Health promotion systems provide certain kinds of health care on a mass-produced basis. These systems can be facilitated by existing organizational arrangements and by centralized management and decision-making—a situation apparently made to order for an industrial setting. Either the company management or the union management can take the lead in bringing such a system into being and making it work. Without this preexisting leadership the problem would be much more difficult.

But the existence of centralized decision-making apparatus brings with it the possibility of several kinds of abuses: paternalism, elitism, subtle forms of coercion, and conflicts of interest. All of these are possible, and are frequently seen. Adversary relationships between company and union can stymie what could otherwise be a very fruitful cooperative venture.

There is no one answer, or simple answer, to these problems. There are a few issues here, however, on which I personally have very strong feelings. Number one, I do not believe that such programs can thrive and reach their maximum effectiveness in the presence of strong adversary relationships between company and union. These programs need to be cooperative ventures in order to have maximum impact.

Number two, although centralized management and decision-making is essential to the success of such programs, there must be grass-roots participation and a voice for all parties. This implies some kind of an advisory board, steering committee, or council that provides a review of policy and a forum for the expression of employee opinion.

Number three, there must be absolute privacy of data. Data are essential to the success of a health behavior change program, and these data must be fully protected. They cannot become available to management, without the

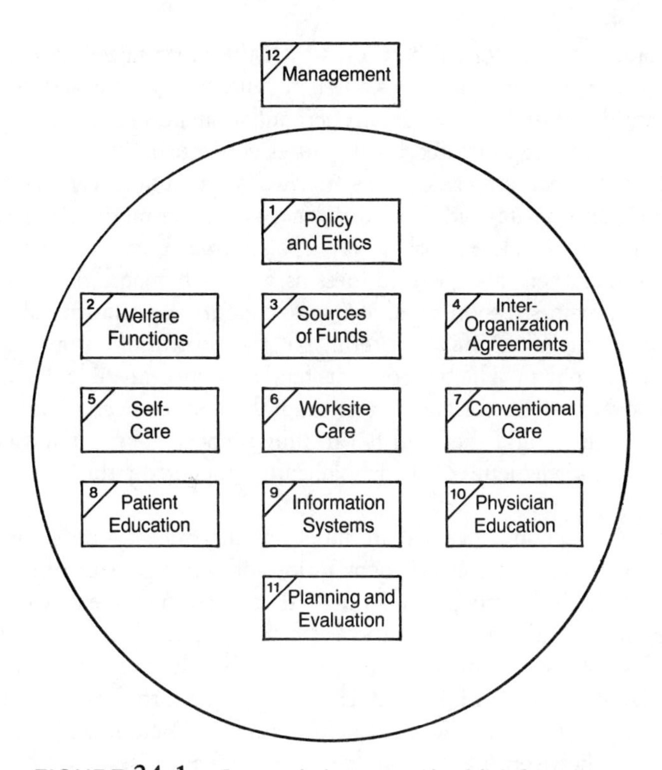

FIGURE 24.1 Principal elements of health behavior change systems at the worksite

specific permission of the individual, and they obviously cannot be mingled with data that the employee is required to provide as a condition of employment.

As noted by McLean and others (4), this last problem is a very difficult one, and may be a stumbling block that many programs will fail to surmount. One key advantage of services provided by a third party is that this party can gather and store data needed for program functions on a private basis.

Welfare Functions

The costs of a health behavior change program at the worksite are in essence a tax on all the employees of the company. The services paid for by these monies go to certain selected groups of employees, depending on the nature of the program. The formal basis on which the program services are distributed, if there is one, constitutes a *welfare function.* These functions deserve more consideration than they usually get, and ultimately should be formalized so that they can be evaluated by all parties concerned.

In somewhat simplistic terms, the welfare functions can tend to emphasize either *company utility* or *social utility.* Concentration of physical examinations on executives is obviously a policy that stresses company utility. I believe that there are good business as well as humanitarian reasons for putting the main stress on *social utility.* For one thing, company utilities are inclined to emphasize the short term, but the control of chronic disease is a long-term problem. The two concepts tend to be incompatible. For another, the employees will see through any program that puts company utilities ahead of social utilities, and they will fail to fully support the program. Thus the most economic element of all—wholehearted support by the employees—is lost.

Some programs have welfare functions that reflect a different fallacy. They seek to maximize cost-benefit ratios without regard to other factors. This policy leads directly to "creaming," that is, taking the easy cases off the top and leaving all the others to take care of themselves. Such programs accomplish almost nothing, since the easy cases would take care of themselves in any event. It is essential to any bona fide program that there be a serious attempt to reach those who need help the most. There must be a division of the funds between the easy and the more difficult cases.

In figuring utility, it is essential to take personal time and effort into account. For the low-grade problems with which we are concerned, these are not negligible factors and in fact are often the dominating factors in determining what action a person will take. There is no logical reason that I can think of why an asymptomatic person should spend an hour or so and $20 to have his blood pressure taken. At the worksite we can in some cases provide not only convenience, but ultra-convenience, and this will be found to be a remarkably powerful tool.

Obviously, one of the reasons there is so much confusion with regard to social welfare functions for health programs is that this aspect of the problem is in a very primitive stage of development. This is an area where the DHHS can and should make a major contribution, since the DHHS is generating the data on which such functions are based, and since such functions are an essential element of program evaluation.

Sources of Funds

One advantage of health behavior change programs at the worksite, if they are really efficient, is that they can make a contribution to cost containment.

In principle, such programs can do better than break even on a cash-flow basis. But there are a number of complications that stand in the way of this being achieved in practice.

Since cost-effectiveness in these programs is not well understood, it is hard to achieve. A health promotion program has to be efficient in each of its elements, well-balanced, and strongly supported by employees to reach the desired goal. Most importantly, it must *actually produce health.* And this aspect of the problem is the least understood of all.

It is not generally realized, but the cash returns from health promotion programs are larger in the life and disability insurance plans than in the health insurance plans (*5*). Preventive care is still care, and most of this care costs money. It appears that stepped up blood pressure control, for example, will actually induce a net increase in health insurance costs (*6, 7*). Therefore small companies will have a hard time getting their money back. Their life and disability plans are not rated on experience, and hence do not pay back when experience improves.

A few insurance companies are beginning to consider providing risk-class ratings for individuals or small companies that would circumvent this difficulty (*8*). This is very much to be encouraged. The DHHS has an obligation to provide a technical basis for such ratings.

Costs and benefits have to be calculated separately for each of the parties involved in a health promotion program. For example, a private health maintenance organization (HMO) serving a company with a stepped-up blood pressure control program will suffer a loss because it is community-rated and will ordinarily provide the extra care at no extra cost. Fee-for-service physicians, on the other hand, tend to reap undue benefits from the extra referral and monitoring that takes place at the worksite (*5*).

Interorganizational Agreements

As just noted, the company itself is not the only party affected economically by a health promotion program. The employees, the union, and the care providers are affected as well. The best results are likely to be achieved when the interests of all parties are recognized and reflected in formal interorganizational agreements that define the roles of all participants.

Working with an HMO can be a particularly powerful way to implement a health promotion program. The company, being at the front end of the process, can initiate a program of stepped-up care by increasing referrals, with or without the agreement of the HMO. But it seems more logical to work out, on a joint basis, a policy that is equitable and agreed on by all parties. Among other things, such agreements can define expected standards of quality and quality control for both the company and the providers.

Self-Care

A key function of any health behavior change program is to encourage *self-care,* that is, increased personal responsibility for one's health and health care. Within proper limits, this is the most cost-effective, and the most *liberating,* care of all. However, there are dangers here, and the effective use of this element calls for really enlightened management.

Self-care cannot simply be a catchword that covers up a laissez-faire policy. Since self-care is *participation,* emphasis on this element puts more demands on the participative aspects of the program. There must be a feedback of information to the employee in terms that he can understand and act on. There must be extra effort to facilitate such actions. It is particularly important that the program be free from managerial bias. The program manager may very properly be a fitness buff, but this bias should not be unilaterally imposed on employees if they are expected to strongly support the self-care aspects of the program.

Again there is something to be said for formal agreements and quid pro quo. For example, management could say to employees: if we have evidence that self-care has reached a certain level, we will start and support a further segment of the health promotion program. Incentives of various kinds can be useful in encouraging the self-care aspects of the program.

Worksite Care

At the heart of any work-based health behavior change program is *worksite care.* There is no substitute for this component, since this is where the efficiencies of care can be the greatest. In most cases, if adversary relationships between the company and the union force such care to be provided offsite, the program loses much of its effectiveness.

Worksite care has more ingredients than are generally recognized. The following are some of the most important:

Risk (health hazard) assessment
Communication to participants
Behavioral assessment
Risk status monitoring
Behavioral status monitoring
Risk reduction planning
Methods of facilitation
Risk reduction actions (interventions)

The first and last of these items have received most of the emphasis to date, and even there a great deal of further development is needed. The items in-between are generally in a still more primitive stage of development.

Risk assessment is not as well developed as is generally supposed. Methods exist, but the validity and accuracy of these methods leave a great deal to be desired. False positive and false negative ratios are high, and these flaws affect the economics of the program. By and large, risk assessment methods used today are not optimal on either an *information-theoretic* or a *decision-theoretic* basis. By this I mean that these conventional techniques do not make use of the potentially available information in an efficient manner, nor do they use information and utility together to form decisions in an efficient manner.

Even in something supposedly as simple as blood pressure screening, the typical approach used produces false negative and false positive ratios that I would consider unacceptable, and are tinged by some superimposed bias as well (9). This is not recognized as a problem because the accuracy of the screening procedures are never evaluated. The DHHS has an important contribution to make in this area, and an obligation to raise the standards of quality in risk assessment.

Communication to participants is a much-neglected area. Poor communication can have iatrogenic as well as adverse economic consequences. Standards of normality are so variable that the same population could be characterized as 5 percent abnormal in one setting and 50 percent abnormal in another. This is chaotic. The following steps are suggested.

First, make sure of laboratory values. If these are not accurate, everything else is a joke. Second, express data in percentile categories against age and sex-weighted normal values, based on large and well-controlled U.S. surveys, such as the National Health Examination Survey. These percentiles are relatively stable and free from personal opinion. Third, define the actual risk ratios implied and convey these to the individual to help him/her form his/her own judgments. Fourth, use normal and abnormal categories that are related in a logical way to risk ratios. Fifth, use graded advisories rather than a binary normal-abnormal categorization. The urgency of the advisory should be proportional to the risk involved.

In relation to health behavior change, behavioral assessments are an important but still relatively undeveloped subject. At present it is possible to assess behaviorally, with modest but still useful degrees of predicting power, certain types proneness to ill-health (e.g., coronary-proneness), to addiction, and to mental health problems, as well as proneness to noncompliance. All of these tendencies are of consequence, for behavior is the primary thing that a health promotion program is concerned with.

Status monitoring of both risk and intervention actions is an important subject that will be discussed under the heading "Information Systems."

Risk reduction planning is a rather complex process that has to take into account for each individual: risk status, risk reduction potential (which may be quite different from risk), resistance (i.e., behavioral problems), and current location in the care system. We are drastically in need of better and more comprehensive algorithms for this planning. In addition, to determine what to emphasize in the program, the program manager has to be concerned with the risk statistics of the target population in various categories. Both individual and program planning put their own demands on the information system.

In both risk assessment and risk reduction, it is important from an economic point of view to adapt the amount of effort to the need. A comprehensive annual physical for all employees would be prohibitively expensive, but it is not needed anyway. In screening, I would personally advocate at least three different frequencies of examination (e.g., three years, one year, and six months) and also three different levels of examination, depending on the level of risk identified at the previous stage.

Interventions also should be applied in steps, with the majority of persons receiving rather low levels of assistance which is all that they need, but with some individuals with more difficult medical or behavioral problems getting special help. This approach is needed to make these programs cost-effective, but by and large the algorithms needed to define who gets what and when have yet to be developed.

It might be interesting to look at the more obvious interventions from an economic point of view. From what we know today, I would estimate that the *cost-benefit* ranking is as follows:

1. Smoking
2. Blood pressure
3. Lipid levels
4. Exercise

Smoking intervention is appreciably better than a break-even proposition; blood pressure and lipid level control are close to the break-even point; and exercise is probably somewhat below the break-even point, considering only the aerobic fitness aspects. Note that a program does not have to break even to be desirable, but that if it does this means that the health effects themselves are achieved at no net cost.

I believe that the ranking of these same interventions according to *popularity* is probably

1. Exercise
2. Blood pressure

3. Lipid levels
4. Smoking

The most notable difference in the two rankings is the switch between exercise and smoking. We are talking about health behavior change at the worksite, but at least until very recently there has been considerable reluctance to take on the most important and cost-effective behavior change of all, smoking cessation.

But perhaps we should not be too theoretical. If this is what people want to do, we should help them do it. Exercise may have more behavioral aspects than are apparent at first glance. It may lead to modifications in the other risk factors—very few marathon runners smoke two packs a day. But since favorable cost-benefit ratios are harder to achieve with exercise interventions, good guidelines are especially important. Perhaps the DHHS should itself put a greater emphasis on exercise as a health intervention.

Conventional Care

In a very few cases almost all aspects of medical care, and not just the outreach aspects, can be provided at the worksite. But in the majority of cases most of the medical care will be provided, after identification and referral, by the conventional care system. There is need here for a well-defined point of interaction and for standardized procedures of information exchange. The worksite has considerable leverage here, being both a source of referral and a source of funds. This leverage should be used to full advantage.

The worksite must provide quality control on its referrals, but then also insist on quality control on the actions taken in the wake of a referral. Because of the complementary nature of worksite care and conventional care the possibilities are good that effective working relationships can be evolved. But this aspect of the problem needs specific attention.

Patient Education

The worksite lends itself to general education, to the use of group esprit, to incentives, and to group incentives more than almost any other environment that can be imagined. These relatively unique aspects of the worksite should be utilized.

The motivation value of games and lotteries should not be overlooked. They can add a little spice to what otherwise can be a very dull subject. The somewhat impulsive person who hates to plan ahead may be exactly the one who is most intrigued by the games, the gambles, and the fanfare.

Perhaps the element that would produce the greatest motivation of all would be the feedback of information showing that the program had been successful. The worksite lends itself to such feedback on a personalized basis, *if the program is indeed effective.* This is the critical challenge.

Information Systems

At present we do not have information systems that are suited to the requirements of health behavior change at the worksite. This is a new form of care, dealing largely with chronic disease, and it requires data systems that are quite different from what are now being used in conventional—episodic and mostly acute—care.

The list of missing ingredients is quite long. We have few really effective risk assessment rules and algorithms. We are almost totally lacking algorithms to define risk reduction potential. We do not have procedures in hand to define the age and sex-weighted percentile categories for each person. Most health promotion data systems that I am aware of lack inverted files (risk by person instead of person by risk) to use as the basis for action planning.

Most, if not all, existing data systems lack action-planning algorithms. They also lack information (including followup data) on behavioral status. It is not surprising, then, that the results achieved are modest.

It seems that we have almost reached the point where information systems for health behavior change at the worksite will have to be computer-based. The problems involved simply do not lend themselves very well to manual manipulations. There is a great deal of room for algorithmic care, since by and large no emergencies and nothing of an acute nature are involved.

The procedures required are more complex than is generally realized, but they are also repetitive and suitable for standardization. Much of what is needed can be handled by microcomputer techniques. Microcomputers will be particularly useful for experiments leading to the development of full-scale systems. This appears to be an area where third-party service agencies could make an important contribution.

Physician Education

Physician education is just as important as patient education. This can be demonstrated statistically. But physician education is practically impossible for small companies to undertake by themselves. Large companies, working through HMOs, are in the best position to promote physician education and feedback.

Planning and Evaluation

Although planning and evaluation are critical to effecting health behavior change at the worksite, they are very weak at present. This is the area where the DHHS can and should make its greatest contribution.

The first need is for models of the health effects of interventions. These must be evolved out of the data from the large DHHS intervention trials. There is no other comparable source of data. Not even the largest company can come close to developing a data base of sufficient size for this purpose. But although the data bases are being developed by the DHSS, in general the models are not, so much more effort must be made in that direction.

The second need is for models of the behavioral effects of interventions. Again the DHHS is the main source, but here more trials are needed. Such trials can be much less expensive than the trials on health effects. The Multiple Risk Factor Intervention Trial *(10)* may not be much more than a smoking cessation trial, when all is said and done, but at least 500 trials of smoking cessation behavior could have been carried out at the same cost.

The third need is for models of operations and costs. These can be developed by many sources, including the industrial concerns involved. The function of the DHHS here should be to collate existing information and make it generally available.

The fourth need is for models of the benefits that result from given health effects—benefits in regard to both social utility and cash flow to the various parties involved. This is a much neglected area because for many years the National Institutes of Health has refused to become involved seriously with economics, and the National Center for Health Services Research has been equally shy of dealing with categorical disease issues. This unfortunate gap badly needs to be bridged. Some of the problems that have to be dealt with in the development of such models are discussed in a special issue of the *Journal of Occupational Medicine (11)*. Discussions of more general modeling issues are provided by Chase *(12)* and Greenberger et al. *(13)*.

Planning and evaluation models, when developed, will be rather complex. One possible method for communicating such models would be to implement them on time-sharing computer systems available to the general public. Ideally, such models would include standard criteria that would allow results at different facilities to be compared on a direct basis. For example, if a measure of the degree of blood pressure control achieved at a given facility were expressed in absolute terms as one number, that number would be closely related to the residual risk from uncontrolled blood pressure at the particular facility.

There are several pitfalls that have to be taken into account in any evaluation of health promotion programs. First, since self-selection is always at work, one has to consider the health of those both in and outside of the specific program. The impact on the whole employee population is the final measure of success. Second, secular trends have to be taken into account. Distinct improvements in health behavior are being accomplished without any formal programs at all. Third, the effect of startup transients needs to be considered in new programs. There will be extra efforts at startup and these will appear to "produce" *ill health* by finding more people with problems. Fourth, impacts of changes in behavior will generally not show up as health effects until two or three years later. The behavioral change programs themselves require a few years to reach their full impact. Finally, spillover effects from one program component to another, and from those in a given program to those not in it, are very desirable but tend to complicate evaluation. Again, for control it is necessary to consider one facility as the unit of experimentation. Obviously, highly detailed guidelines will be needed, and the DHHS is the only agency that can fill the gap in this area.

REFERENCES

1. Chadwick, J. H. Costs and cash benefits of heart disease programs at work. Supplement to Guidelines no. 1035, SRI International Business Intelligence Program, November 1978.
2. U.S. Bureau of the Census. *Statistical Abstract of the United States.* 98th annual ed., 1977.
3. Chadwick, J. H. Heart disease control programs. Guidelines no. 1035. SRI International Business Intelligence Program, September 1978.
4. McLean, A. Management of occupational health records. *Journal of Occupational Medicine* 18:530–533 (1976).
5. Warner, K. E. Health maintenance insurance: Toward an optimal HMO. *Policy Sciences* 10:121–131 (1978).
6. Stokes, J., and Carmichael, D. C. A cost-benefit analysis of model hypertension control. Paper published by the National High Blood Pressure Education Program of the National Heart and Lung Institute, National Institutes of Health, Bethesda, Md., May 1975.
7. Weinstein, M. C., and Stason, W. B. *Hypertension: A policy perspective.* Cambridge, Mass.: Harvard University Press, 1976.
8. Kotz, H. J., and Fielding, J. E., eds. Health, education and promotion: Agenda for the eighties. Summary report of an insurance industry conference on health

education and promotion, Atlanta, Georgia, March 16–18, 1980 (sponsored by the Health Insurance Association of America).
9. Chadwick, J. H., and Black, G. W. Rates of error in blood pressure screening. Paper presented at the National Conference on High Blood Pressure Control, Los Angeles, California, April 2–4, 1978.
10. MRFIT Research Group. The Multiple Risk Factor Intervention Trial (MRFIT): Results after four years of intervention. Paper presented at the Conference on Epidemiology, Washington, D.C., March 27–29, 1981 (sponsored by the Council on Epidemiology of the American Heart Association).
11. Hughes, J. P., ed. Cost-effectiveness of occupational health programs. *Journal of Occupational Medicine* 16:153–186 (1974).
12. Chase, S. B., Jr., ed. *Problems in public expenditure analysis.* Washington, D.C.: Brookings Institution, 1968.
13. Greenberger, M.; Crenson, M.; and Crissey, B. *Models in the policy process.* New York: Russell Sage Foundation, 1976.

25

Hypertension Control Programs in Occupational Settings

Michael Alderman, M.D., Lawrence W. Green, Dr.P.H., and Brian S. Flynn, Sc.D.

Hypertension represents an area of health where a modest investment ought to yield a great benefit, if indeed an ounce of prevention is worth a pound of cure. Cardiovascular disease, including 600,000 heart attack deaths and 60,000 stroke deaths, was the leading cause of death and disability in the nation in 1976 *(1)*. In that year, these cardiovascular diseases consumed, in direct and indirect costs, some $50 billion, or 20 percent of all health-related expenditures.

Although the causes of the major cardiovascular diseases—heart attack and stroke—are not entirely understood, certain conditions or characteristics are known to increase the likelihood of their occurrence. The major known

This article was first published in *Public Health Reports* (March–April 1980), pages 158–163, and appears here in slightly revised form. It was supported in part by DHEW research training grant 1-T32-H10710-03 to Johns Hopkins University.

Dr. Alderman is Associate Professor of Medicine and Public Health at Cornell University Medical College. Dr. Green, formerly Director of the Office of Health Information and Health Promotion, is now a visiting lecturer at Harvard Medical School and has been appointed Director of the Center for Health Promotion Research and Development at the University of Texas Health Science Center. Dr. Flynn, formerly a doctoral candidate in the Division of Health Education at Johns Hopkins University, is now Associate Director of the Vermont Lung Center, University of Vermont Medical School.

risk factors are a family history of premature vascular disease, smoking, hyperlipidemia (high blood fats), and hypertension. Of these, hypertension has the strongest association with subsequent cardiovascular catastrophes, such as heart attack and stroke, and is most responsive to available interventions.

During the past generation, substantial progress has been made in defining the natural history of high blood pressure and in developing effective means of treatment. Blood pressure elevation usually is first detected during the fourth or early fifth decade of life. Although this disease usually has no symptoms, early detection and treatment can reduce the incidence of cardiovascular death and disability.

Unfortunately, the best current information is that the majority of those with high blood pressure are not receiving effective therapy. A national antihypertension campaign, however, has altered substantially the nature of the problem. In contrast to the situation of only a decade ago when perhaps one-half of all hypertensives were undetected, now more than 80 percent of those afflicted are aware of their conditions and fully two-thirds are receiving treatment. The glaring defect in the current situation is the widespread failure of the health care system to achieve and maintain long-term blood pressure control after patients begin treatment. As many as half of the patients who begin treatment for hypertension do not remain under care or do not adhere adequately to therapeutic recommendations.

In an effort to bridge the gap between the technical potential and the actual achievement of blood pressure control, a variety of structural and educational strategies have been developed and evaluated both in occupational and other settings. In this paper, we review (*a*) several types of hypertension control programs that have been tested in occupational settings and (*b*) the results of a series of studies conducted in occupational and other settings to determine effective educational methods for helping patients to maintain their hypertension medication regimen.

STRUCTURES FOR CONTROL PROGRAMS

Hypertension control activities in occupational settings have been underway for a number of years. Medical, economic, social, and logistical imperatives have produced several approaches to achieving blood pressure control. Because the knowledge that blood pressure reduction can lessen the risk of cardiovascular disease is so recent and adequate implementation and analysis of health care delivery programs require so much time, information about blood pressure control activities at workplaces is incomplete. Nevertheless, a number of well-documented studies have accumulated sufficient evidence to show that control efforts in occupational settings offer considerable promise of contributing to workers' wellness.

Two major categories of efforts to promote blood pressure control at the worksite are (a) detection of hypertensive employees at the worksite and referring them to community resources for continuing treatment—a systematic followup program has been an integral component of these programs—and (b) provision of antihypertensive therapy at or near the worksite. In several instances, the two patterns have been combined.

Detection, Referral, and Follow-up

Chicago Heart Association Project

The workplace was perhaps first used as a means of casefinding and referral in the mid-1960s. The Chicago Heart Association, in collaboration with 81 industries, systematically screened some 37,714 (55 percent) eligible employees *(2)*. Hypertension was detected in 19 percent of these employees, but at the time of screening it was being controlled satisfactorily in fewer than 15 percent. Although 65 percent of these hypertensives actually saw a referral physician, follow-up for the group as a whole revealed that they had not achieved significant improvement in blood pressure control after five years.

The directors of the Chicago project concluded that their initial approach—merely advising screenees with elevated blood pressure levels to see a physician—was not sufficient to produce and maintain the necessary long-term therapy. Therefore, in 1969, they began another approach to detection and referral of hypertensive employees. This approach substantially increased communication between hypertensive workers and staff of the hypertension control program because it included additional contact with those with elevated blood pressure readings at the first screening. Laboratory tests were performed, and the results were given to the hypertensive employees for transmission to their physicians. In addition, a health educator spoke with each of these employees, and five one-hour classes on cardiovascular disease were offered to the affected workers at their workplaces.

Unfortunately, the response to this second strategy was as unimpressive as the response to the first approach. Fewer than half of the hypertensive employees actually visited a physician, and two years later only one-half of those who sought care had achieved and maintained control of their blood pressure.

The results of the Chicago project suggest that although workplace screening might be a convenient and economical way to detect hypertension, it does not necessarily lead to either participation in treatment or improved control of blood pressure. Simple referral to customary sources of care does not seem to effect the change in behavior necessary to achieve these goals.

Michigan Worker Health Program

Investigators at the University of Michigan developed a worksite hypertension control program based on a coordinated detection, referral, and follow-up campaign *(3)*. This program, carried out in collaboration with labor unions and management in several work settings, included maintaining contact with both the worker and his or her physician after screening revealed elevated blood pressure. As much effort as necessary to produce effective referral was expended. Follow-up efforts were conducted primarily by mail and telephone. As a result, 88 percent of all identified hypertensives actually consulted a physician. Thereafter, contact with both patients and physicians was maintained at semiannual intervals. With follow-up data available for up to two years, more than 80 percent of the successfully referred employees maintained satisfactory blood pressure control.

The factor that distinguishes the Michigan program from less successful programs is the project staff's systematic and diligent follow-up of both patients and physicians. It seems that this approach is more likely to produce blood pressure control than simple referral without vigorous follow-up.

The Burlington Industries Program

A third approach to detection, referral, and follow-up is represented by a program initiated by Burlington Industries in 1974 *(4)*. A well-planned educational program for management and workers resulted in 100 percent participation in the screening component of this program. Appointments were made with a physician chosen by each worker having an elevated blood pressure. The worker's physicians received a letter containing a record of the blood pressure readings and an offer to provide the worker with blood pressure checks, education, and other services at the worksite in cooperation with the physician. The workers who participated thus continued as patients of their personal physicians for treatment of hypertension, but received supplementary services from the worksite hypertension program to monitor their blood pressure and to help them continue in treatment and maintain their medication regimens. About one-half of the referred employees were approved by their physicians for participation in the cooperative care component of this program. No data on blood pressure control were available from this pilot study, but long-term contact with hypertensive employees and communication between the worksite program staff and community physicians have been found in other studies, such as the Michigan Worker Health Program, to be effective approaches. The Burlington study demonstrated the feasibility of close cooperation between community physicians, hypertensive employees, and staffs of worksite hypertension control programs.

Occupationally Based Treatment

Several investigators have designed comprehensive programs for the detection and treatment of hypertension in the occupational setting. One such program, developed by Cornell University Medical College for the United Storeworkers Union in New York City, detected and treated hypertensive employees at Gimbels and Bloomingdale's department stores (5). Since 1973, over 20,200 employees have been screened. Of the approximately 3,000 who have been identified as hypertensive, some 2,200 are now receiving care at one of 19 union-provided treatment locations. The program uses a health team approach: a nurse, supervised by a physician, provides care according to a systematic protocol. There is no direct cost to the patient for visits, drugs, or laboratory tests. In 1978, total program costs, including laboratory studies and drugs, was $194.77 per person per year. Patients' adherence to treatment has been high, with attrition amounting to less than 10 percent per year. Satisfactory blood pressure control has been achieved and maintained by 80 percent of the program's active patients. Preliminary data suggest that absenteeism and hospitalization have declined for treated patients.

Recent studies have demonstrated that equally satisfactory results can be achieved for employee groups treated at an offsite union health center clinic. This experience suggests that a rigid therapeutic approach, reliance on a health team, removal of personal financial impediments, emphasis on patient participation in the treatment process, and the provision of all services within a socially cohesive institution are individually or together more important determinants of outcome than the physical location of the treatment facility.

Programs in Progress

The projects just described, and other similar ones, have stimulated the establishment of a variety of formal demonstration and evaluation projects designed to measure the relative merit of these various approaches. The University of Michigan, in collaboration with the Ford Motor Company, has established three distinct intervention programs that include two methods for detection and referral and one for onsite treatment at three automotive plants. No intervention is planned for a fourth plant so that results can be compared with those of the experimental locations. The Westinghouse Electric Company recently was awarded a contract by the Department of Health, Education, and Welfare (now the Department of Health and Human Services) to undertake a prospective analysis of different approaches to antihypertensive therapy in various plants. The University of Maryland also has been awarded a grant by DHHS to conduct a study of blood pressure control for state employees. It is expected that prospective collection of economic and medical data will permit accurate assessment of the relative merits of the various approaches being tested.

Understandably, insurance companies also have become interested in the development of programs to encourage blood pressure control activities in industry. The Blue Cross Association, with support from the National Heart, Lung, and Blood Institute, has developed a strategy through which it hopes to demonstrate that industrial blood pressure control activities can be stimulated by the educational efforts of local company representatives. The Massachusetts Mutual Life Insurance Company in Springfield has instituted an occupationally sponsored program in which education, detection, and follow-up of hypertensives is carried out at the worksite and community physicians provide treatment. An additional feature of this program has been the assumption of full costs by the company. Physicians provide information on patient status through the billing process so that vigorous patient follow-up is assured and accurate cost-benefit analysis is feasible.

A variety of worksite-based programs to detect, refer, follow up, and treat hypertension among employees have been initiated in different kinds of occupational settings. In many cases their experience has been long enough to permit some outcome analysis. The results have been sufficiently encouraging to justify their expansion. Studies are now addressing the issue of cost in occupational programs for blood pressure control *(6)*. Logan and co-workers have, in fact, reported that a worksite blood pressure treatment program is more cost-effective than community-based care *(7)*.

EDUCATIONAL STRATEGIES

The National Heart, Lung, and Blood Institute has sponsored 11 studies to test various strategies for improving compliance of patients with blood pressure control *(8, 9)*. The principal investigators of these studies met several times over the course of their three-year grants. In sharing and exchanging their experiences and findings, the investigators arrived at several conclusions that are particularly relevant for occupational health applications. (The results of most of these studies have not yet been published.)

Health Care Providers

Increased Contact Time

In all 11 studies it was found that short-term improvement in blood pressure control can be achieved by almost any intervention that provides more time for discussion between a health care provider and a hypertensive patient. This finding suggests that periodic blood pressure counseling, even without a highly structured educational or behavioral intervention, could possibly be a cost-beneficial activity of worksite health programs. We offer this possibility with trepidation in view of the various types and designs of the studies on which it is based. Nevertheless, there are theoretical explanations for the phenomenon of reduced blood pressure resulting from such generalized intervention. Most such explanations can be categorized as experimental effects or placebo effects *(10)*. Since all the studies from which this observation was drawn were experimental, reduced blood pressure could result from the patients knowing that they were being observed and therefore taking greater care to bring their pressure down by compliance during the period of the study.

Some of the studies, however, used resident staff of the clinical setting rather than research staff to conduct the interventions, thereby minimizing the possibility that the patients would respond on the basis of knowing they were part of an experiment. The second explanation is that the mere process of being under more intensive care and observation resulted in a reduction of blood pressure without necessarily increasing compliance or other behavioral changes. Such a placebo effect is well documented in relation to a wide range of phenomena, including blood pressure.

Increased Number of Contacts

A second observation from many of the studies was that the frequency and continuity of contact between patient and health care provider resulted in greater blood pressure control. This observation is similar but it is in partial contrast to the first observation, which was concerned more with amount of time and intensity of each contact as opposed to the number and variety of contacts over time. The Johns Hopkins data revealed an effect from the number of contacts, but it was not the number of contacts alone that accounted for this effect; rather, it was the combination of content and contact resulting in more opportunities for and types of repetition and reinforcement of behavioral changes *(11)*. In the University of California study in Oakland, a similar contact-content difference was found between home visits and regular contacts with clinic personnel *(12, 13)*. Again, the advantage of the worksite is obvious. With regular and appropriately spaced contacts at the worksite, the schedule of behavioral change can be paced and tailored to the abilities and motivation of the worker and reinforced over time by subsequent contacts.

Active Patient Participation

A third generalization drawn from these studies suggests another level of intervention. The 11 studies appear to have achieved varying levels of blood pressure control, depending on the extent to which patients were actively rather than passively involved in setting goals for their own blood pressure control or behavioral change. Some of the most dramatic effects were achieved, for example, in small studies in which each patient literally contracted with the health care provider for behavioral and blood pressure achievements that would be rewarded by the provider with tangible goods such as trading stamps, a book, or a ticket to a sporting event (*14–16*). In the Johns Hopkins study in Baltimore it was noted that of three kinds of intervention the one that required the greatest amount of participation by the patients, as distinct from provider-initiated contact, resulted in the greatest reduction in blood pressure (*17*).

Social Support

A fourth common observation in most of the 11 studies was that the involvement of a significant other person in addition to the patient and the provider was helpful in reducing blood pressure at least temporarily. Two of the University of Michigan studies were designed specifically to test the hypothesis that the enlistment of a partner in blood pressure control would have this effect. Perceived social support was significantly increased, and preliminary analyses suggested that blood pressure control corresponded with social support (*18–21*). The North Carolina study at Chapel Hill successfully mobilized the effect of social support through family members or friends who were trained to measure the patient's blood pressure and by home visits by a pharmacist or nurse (*22*). The Johns Hopkins study also stimulated such social support through home visits (*23*). The potential of the worksite as a more convenient place than the home to mobilize social support for blood pressure control suggests an even greater potential for achieving this effect with co-workers in occupational health programs.

Self-monitoring of Blood Pressure

Some of the studies found an added effect when patients were given the opportunity to monitor their blood pressure and to keep records of changes (*15, 20, 22, 24, 25*). This effect is believed to operate through another form of direct feedback and reinforcement. When the changes in blood pressure are more visible and accessible to the patients, these patients can adjust a variety of life-style habits according to the trends and changes in their blood pressure. Some of these changes may not have been so obvious to the health care provider, much less the patient, without such direct feedback. Thus, in the workplace the development of self-monitoring skills with accessible resources could be easier than in most medical sites or homes *(24)*.

COMMENTS

From the work reviewed, it is clear that hypertension control programs that are focused primarily on detection of elevated blood pressure are unlikely to achieve benefits sufficient to justify their costs. Programs that include efforts to ensure completion of referrals may achieve this intermediate objective, but they do not address the problem of patients' failure to stay in treatment or to adhere to medication regimens—the additional steps necessary to achieve blood pressure control.

Programs that include some form of long-term contact with hypertensive employees appear to be successful in achieving blood pressure control. The success of these programs is enhanced by contact with the hypertensive worker's personal physician or by the provision of services for the patient at the worksite, or both. The findings on the effectiveness of increased intensity, variety, and number of contacts and on the effectiveness of social support are consistent in studies of program structures and of patient education strategies. One of these studies has now established a 57 percent reduction in mortality five years after intervention *(26)*.

The experiences of the programs reviewed afford some warnings and some encouragement. The cautionary conclusions come from the poor results and short-lived effects of simplistic screening, referral, or educational programs that have no follow-up and reinforcement over time. The encouraging signs come from worksite programs that are more intensive—they involve hypertensive employees in their own care and vigorously maintain contact with these employees over time. Additional encouragement is drawn from the results of recent studies of educational interventions in clinical settings. Their methods could be used and probably their effective results could be achieved in occupational settings.

REFERENCES

1. American Heart Association. *Heart facts.* Dallas: National Center, 1980, p. 11.
2. Schoenberger, J. A. Heart disease in industry: The Chicago Heart Association

Project. In *High blood pressure control in the work setting: Issues, models, resources.* Proceedings of the National Conference, Washington, D.C., October 14, 1976. West Point, Pa.: Merck, Sharpe and Dohme, n.d.
3. Foot, A., and Erfurt, J. A model system for high blood pressure control in the work setting. In *High blood pressure control in the work setting: Issues, models, resources.* Proceedings of the National Conference, Washington, D.C., October 14, 1976. West Point, Pa.: Merck, Sharpe and Dohme, n.d.
4. Murphy, A. F. The Burlington Industries industrial hypertension program. In *High blood pressure control in the work setting: Issues, models, resources.* Proceedings of the National Conference, Washington, D.C., October 14, 1976. West Point, Pa.: Merck, Sharpe and Dohme, n.d.
5. Alderman, M. H. Detection and treatment of high blood pressure at the work place. In *High blood pressure control in the work setting: Issues, models, resources.* Proceedings of the National Conference, Washington, D.C., October 14, 1976. West Point, Pa.: Merck, Sharpe and Dohme, n.d.
6. Ruchlin, H. S., and Alderman, M. H. Cost of hypertension control at the workplace. *Journal of Occupational Medicine* 22:795–800 (1980).
7. Logan, A. G., et al. Cost-effectiveness of a worksite hypertension treatment program. *Hypertension* 3:211–218 (1981).
8. McGill, A. M. A National High Blood Pressure Education Research Program; Abstracts of papers presented at the First International Congress on Patient Counseling. *Patient Counseling and Health Education* 1:35 (1978).
9. National High Blood Pressure Education Research Program. Announcement, NIH guide for grants and contracts. Bethesda, Md., 1973, vol. 2, pp. 3–4.
10. Green, L. W. Evaluation and measurement: Some dilemmas for health education. *American Journal of Public Health* 67:155–161 (February 1977).
11. Chwalow, A. J., et al. Effects of the multiplicity of interventions on the compliance of hypertensive patients with medical regimens in an inner city population. *Preventive Medicine* 7:51 (March 1978).
12. Syme, S. L. Hypertension education program in a low income community. Final report on NIH grant No. HL16959, 1976 (National High Blood Pressure Education Program).
13. Fisher, A. A.; Hussein, C. A.; and Syme, S. Congruence between self-reported and staff perception of compliance: Hypertension management. School of Public Health, University of California, Berkeley, 1977.
14. Brucker, C. Assuring patient compliance by health care contracts. Final summary report on NIH grant No. HL17230, 1977 (National High Blood Pressure Education Program).
15. Steckel, S. B., and Swain, M. A. Contracting with patients to improve compliance. *Hospitals* 51:81–84 (1977).
16. Swain, M. A. Experimental interventions to promote health among hypertensives. Paper presented at American Psychological Association annual convention, Toronto, Canada, 1978.
17. Levine, D. M.; Green, L. W.; and Deeds, S. G. Health education for hypertensive patients. *Journal of the American Medical Association* 241:1700–1703 (1979).

18. Caplan, R. D., et al. *Adhering to medical regimens: Pilot experiments in patient education and social support.* Institute for Social Research, Ann Arbor, Michigan, 1976.
19. Flowers, R. V. Effects of social support on adherence to therapeutic regimens. Doctoral dissertation. University of Michigan, 1978.
20. Kirsht, J. P., and Rosenstock, I. M. Patient adherence to antihypertensive medical regimens. *Journal of Community Health* 3:115–124 (Winter 1977).
21. Caplan, R. D., Research Center for Group Dynamics, Institute for Social Research, Ann Arbor, Mich. Personal communication (regarding NIH grant No. HL18418), 1978.
22. Earp, J. L., and Ory, M. G. The effects of social support and health professional home visits on patient adherencc to thc hypertension regimens. Abstract based on year three progress report on NIH grant No. HL18414, 1978 (National High Blood Pressure Education Program).
23. Fass, M. F.; Green, L. W.; and Levine, D. M. The effect of family education on adherence to antihypertensive regimens. Paper presented at National High Blood Pressure Control Conference, Washington, D.C., 1977.
24. Chadwick, J. H.; Chesney, M. A.; and Jordan, S. C. Blood pressure education in an industrial setting: Notes for a progress report to the National High Blood Pressure Education Research Program. Stanford Research Institute, Menlo Park, Calif., 1977.
25. Solomon, H. S. Hypertension—education models to improve adherence. Summary progress report on NIH grant No. HL18423–03 (National High Blood Pressure Education Program). Peter Bent Brigham Hospital, Boston, 1977.
26. Morisky, D. E.; Levine, D. M.; Green, L. W. Five-year blood pressure control and mortality following health education for hypertensive patients. Paper presented at the American Public Health Association, Los Angeles, California, 1981.

26

Weight Control and Nutrition Education Programs in Occupational Settings

John P. Foreyt, Ph.D., Lynne W. Scott, M.A., R.D., and Antonio M. Gotto, M.D.

Malnutrition has become a major health problem in the United States. Both underconsumption and overconsumption of food contribute significantly to death and disease. Coronary heart disease, hypertension, obesity, diabetes, and other conditions related to food account for well over half of all deaths each year.

The U.S. diet has become overladen with saturated fat, cholesterol, sugar, and salt and deficient in complex carbohydrates, such as grains, vegetables, and fruits.

Because of changes in eating habits since the turn of the century, several authoritative bodies, including the Council on Food and Nutrition of the American Medical Association and the Food and Nutrition Board of the National Academy of Sciences *(1)*, have recommended that persons at risk of developing heart disease follow a diet aimed at maintaining ideal body

---

This article was first published in *Public Health Reports* (March–April 1980), pages 127–136, and appears here in slightly revised form.

Dr. Foreyt is Director, and Ms. Scott is Chief Dietitian of the Diet Modification Clinic, National Heart and Blood Vessel Research and Demonstration Center, Baylor College of Medicine and the Methodist Hospital. Dr. Gotto is Professor and Chairman of Medicine, Baylor College of Medicine and the Methodist Hospital.

weight and lowering plasma cholesterol. The Inter-Society Commission for Heart Disease Resources *(2)* has proposed that the entire American population follow these recommendations. The Senate's Select Committee on Nutrition and Human Needs has suggested specific dietary goals for Americans:

1. To avoid overweight, consume only as much energy (calories) as is expended; if overweight, decrease energy intake and increase energy expenditure.
2. Increase the consumption of complex carbohydrates and "naturally occurring" sugars from about 28 percent of energy intake to about 48 percent of energy intake.
3. Reduce the consumption of refined and processed sugars by about 45 percent to account for about 10 percent of total energy intake.
4. Reduce overall fat consumption from approximately 40 percent to about 30 percent of energy intake.
5. Reduce saturated fat consumption to account for about 10 percent of total energy intake; and balance that with polyunsaturated and mono-unsaturated fats, which each should account for about 10 percent of energy intake.
6. Reduce cholesterol consumption to about 300 mg a day.
7. Limit the intake of sodium by reducing the intake of salt to about 5 grams a day *(3)*.

If followed, these suggestions would have Americans eating less red meat, eggs, sugar, salt, and foods high in saturated fat, and more whole grains, vegetables, and fruits.

The worksite offers a unique opportunity for implementing suggestions to improve the U.S. diet. Unfortunately, research has consistently shown that good advice on food consumption is rarely followed. Programs for weight loss, for example, have been notoriously ineffective *(4)*. It has been said that the chances of an obese person losing 40 pounds and maintaining that loss for five years are about the same as those of a person recovering from cancer of the stomach. Combinations of low-calorie diets, exercises, and medication have been unsuccessful in the great majority of cases. Although some people may lose a few pounds while actively following a treatment program, none of these three treatment techniques has been shown to result in significant long-term success for most obese persons.

In the last 10 years, however, a new strategy has been aimed specifically at teaching people not only what to eat but how best to adhere to a recommended diet. A number of techniques directed at behavior change have shown some recent success in altering habits that traditionally have been extremely difficult to reform. In this paper we will review both the specific techniques and the programs in which they have been used, especially programs that have been carried out at the worksite, as well as programs for obese persons and those aimed at changing diet in general.

BEHAVIORAL TECHNIQUES

The basic premise underlying behavioral treatment programs is that dietary changes require substantial changes in eating habits. Poor diet, including underconsumption and overconsumption of foods, is seen as stemming primarily from environmental sources, not from an underlying pathological condition. The goal of these programs is to teach people how to achieve self-control over their eating behavior. Treatment begins with a detailed investigation of a person's dietary behavior through self-monitoring; the person writes down all important events that occur before the ingestion of food and tend to elicit eating, the actual act of eating, and the events that occur after eating that seem to maintain the intake of certain foods. The person is then taught specific techniques, called stimulus control techniques, to help change the events that occur before eating; the act of eating, such as slowing down consumption; and events after eating (contingency management techniques). Most behavioral programs use all of these techniques, and each has been investigated experimentally.

Self-monitoring

Essentially all behaviorally oriented treatment programs for dietary change use some form of self-monitoring that consists of two steps: observation of one's dietary behaviors and recording those observations. The person is asked to be aware of and write down his dietary behavior during the period that he is receiving treatment. He records in a notebook all food and drink ingested, including the amounts consumed, method of preparation, the time of day the food was taken, the place, other people present, feelings just before eating, and other important information. Calories or food exchanges may also be recorded. All information is written down immediately after consuming food or drink. The chart is an example of a daily dietary record. (See Figure 26.1 on page 177.)

Self-monitoring is useful. First, it gives the dieter and the person working with him a record of his dietary life-style. It therefore will help clarify what life-style changes are needed. Second, the act of writing down food intake seems to have a positive effect on the diet itself, at least at the beginning of a program. A person may choose not to eat a super-rich dessert because he does not want others to see his failure or lack of will power.

The technique has been investigated experimentally a number of times *(5–14)*. Most group leaders feel that self-monitoring is important because a record of what one is doing is needed. Some also find that it is helpful in changing behavior per se, although the long-term evidence is not strong. Nonetheless, it has become a necessary component of treatment, although participants often find it inconvenient to fill out the forms, and the task seems boring and monotonous. Once a person keeps such a record for about three weeks, complaints about it tend to lessen. Most seem to become aware of the importance of this dietary record to themselves and their treatment.

Body weight is self-monitored in many programs. Each person is given a graph (body weight makes up the vertical scale and the days the horizontal one) and asked to post it on the wall above the scale. The person is asked to weigh himself on a regular schedule and record the results, an act that makes the person aware each day exactly how he is doing. Individuals may also be asked to monitor their energy expenditures.

Stimulus Control Techniques

After monitoring himself for a week or two, a person will begin to see patterns in his dietary life-style. The physician, dietitian, or other helping person will also see cues that precede eating. For example, he may be eating two doughnuts each morning with a cup of coffee because he stops off on the way to work and brings in a dozen for himself and co-workers. Vending machines stocked with high-fat sandwiches or sugar-laden junk foods may be near the assembly line or the employee's desk. In the evening, the Monday night football game or movie of the week may lead to beer drinking and snacking. Bringing office work home may elicit consumption of certain foods in the evening. Reading office reports or other material oftentimes tends to be tied to snacking. Stimulus control techniques aid the person by helping him rearrange his environment to reduce the chances of inappropriate eating. For example, if it is difficult to drive by the doughnut shop without stopping, the employee might select another route to work in the morning. Vending machines can be stocked with wholesome foods or moved away from the work area. If television watching or reading office reports at home is frequently associated with inappropriate snacking, turning off the set or staying away from the reports while eating may help.

Many people eat in several places at work and at home. To decrease the cues that lead to eating, they may be asked to eat only at one place at work, such as the company cafeteria, and at home, usually sitting at the dining table. Eating is scheduled only at certain regular times both at work

Name Tom Smith

I.D. No. ______ Date Dec. 4

Write ONE food on each line.

7 Meat
2 Dairy
6 Fat
4 Fruit
7 Bread

Time	Place	Amount	Food—How Prepared	Food Exch.
7:30a	kitchen	4 oz	orange juice	1 Frt
	table	3/4 cup	cornflakes with	1 Brd
		1 cup	skim milk & sugar sub.	1 Dr
		1 slice	toast	1 Brd
		1 tsp	margarine	1 Fat
10:30a	lounge	1	fresh peach	1 Frt
12:00p	company	3 oz	baked chicken (no skin)	3 Mt
	cafeteria	1/2 cup	rice	1 Brd
		1/2 cup	green beans with	free
		1/2 tsp	margarine	1/2 Fat
		1 cup	tossed salad with	free
			lemon juice	free
		1	hot roll	1 Brd
		1 tsp	margarine	1 Fat
		1/2 cup	fresh fruit compote	1 Frt
3:00p	vending	1 cup	plain yogurt (low fat)	1 Dr+
	machine			
7:00p	dining	4 oz	broiled steak (no fat)	4 Mt
	room	1 med	baked potato w/ chives	1 Brd
			broccoli with	free
		1/2 tsp	margarine	1/2 Fat
		2 cups	fresh vegetable salad	free
		4 tsp	with French dressing	2 Fat
		1 slice	French bread	1 Brd
		1 cup	fresh strawberries	1 Frt

FIGURE 26.1 Food record

and home. New cues for eating may also be arranged. For example, eating might be allowed only while sitting in a special chair in the company cafeteria or only in the presence of a certain tablecloth, or even from a specially colored plate. The same techniques can also be employed at home. Their purpose is to help break poor dietary habits by restructuring, at least for a while, the employee's worksite and home environments.

These stimulus control techniques are helpful components of dietary treatment programs. Of course, to make the best use of these principles, the dieter and the person helping him must be creative in designing and individualizing the most appropriate procedures for the worksite and home.

Slowing the Act of Eating

It is often helpful for an overweight person to eat slowly. By doing so, the person may better experience the physiological effects of satiety. Since it takes approximately 20 minutes for satiety to occur, the fast eater can ingest a large number of calories without feeling satiated. There are a number of techniques aimed at teaching a person to slow down. He is told to take a bite of food, lay down his fork and knife, put his hands in his lap, and chew slowly before swallowing. For foods not requiring utensils, such as a lunchtime sandwich, the procedure is to lift the sandwich, take a bite, put the sandwich down, hands in lap, chew thoroughly, and swallow. A really fast eater will have great difficulty following such a technique because he forgets after a few bites. To help remind him, he might place a little sign on a 3- by 5-inch card next to his plate: "Lay down utensils between bites" or "Slow down." Eating slowly often makes the experience more pleasurable and the food taste better. The person may also find himself eating less. When an employee is given only a half hour for lunch, he should spend the 30 minutes eating, not shopping, reading the newspaper, running errands, and eating. Restructuring one's time is obviously a crucial element in behavioral treatment.

Contingency Management Techniques

A person needs to be rewarded for changing his diet, and reward systems are built into many programs. For example, an employee may receive five points or tokens a day for keeping food records, eating only in the cafeteria at appropriate intervals, and eating all foods slowly. When 25 points are earned, the person can reward himself with something special (not food). He is also praised for his good work.

Assignments are usually written as a formal "contract" that the employee signs. Contracts make explicit what a person is expected to do, and they may serve as a motivator to keep him in the program. They are especially useful in weight control programs.

WEIGHT LOSS PROGRAMS

Published articles on the behavioral treatment of obesity have been reviewed by a number of authors *(15–27)*. To date, many of the behavioral programs have been conducted in colleges and universities, usually for a group of mildly overweight undergraduate females or people from the surrounding community. These studies have generally shown relatively modest weight losses for the participants. To illustrate the results, we combined the data from 11 published studies *(8, 9, 28–36)*, selected because they present pre-, post-, and follow-up data. A total of 501 persons received some form of behavior modification, 157 received nonbehavioral treatment, usually called "supportive counseling," and 74 persons served as no treatment controls.

The weighted means for the 501 persons who received behavior modification training were 174 pounds pretreatment, 167 pounds after 8 weeks of treatment, and 167 pounds after an 18-week follow-up. The 157 people who received some other form of treatment had an average pretreatment weight of 172 pounds, a weight of 170 pounds after 12 weeks of treatment, and a weight of 174 pounds after a 28-week follow-up. The 74 no-treatment control persons had a weighted mean of 159 pounds; 67 had weighted means of 159 pounds after 9 weeks and 156 pounds (for 42 persons) after 4 weeks. Based on these results, which include a representative sampling of the experimental literature, it appears that with mildly overweight populations, behavioral treatment shows greater posttreatment weight losses than groups not treated behaviorally.

Recently, a number of studies have been carried out with chronically obese individuals. The two largest studies reported to date have been conducted at the Medical University of South Carolina *(37)* and at the Stanford University Eating Disorders Clinic *(38)*.

Average weight of the 165 patients at the start of the 18-month treatment at the Medical University of South Carolina was 185 pounds; they had been overweight for an average of 20 years. At posttreatment, 71 patients had lost less than 11 pounds, 46 had lost between 11 and 20 pounds, 38 more than 20 pounds, and 10 more than 40 pounds. One year later, 56 of these patients were located; 39 had regained their losses, 6 had maintained their original losses, and 11 had continued to lose weight, an average of 9 pounds.

The Stanford researchers reported data on the first 125 clients treated at the clinic. Pretreatment average weights were 265 pounds for 23 males and 206 pounds for 102 females. At posttreatment, 65 clients had lost less than 11 pounds (16 of these had gained weight), 39 had lost between 11 and 20 pounds, 14 between 21 and 30 pounds. 3 between 31 and 40 pounds, and 4 more than 40 pounds. Average weight loss was about 11 pounds. One-year posttreatment data were also reported for 88 clients. Their posttreatment weight loss was an average of 13 pounds, maintained at the follow-up.

The few weight control programs in industry are part of general exercise programs. The weight loss effort may consist of a lecture or two by the company nurse. Some firms have contracted with organizations such as the local health department or a group like the American Diabetes Association or a commercial venture such as Weight Watchers. A detailed example of a contract with an outside group is presented subsequently. Among the few firms that have tried weight loss programs for their employees are the Ford Motor Company, General Foods, and Kimberly-Clark.

Ford Motor Company, Dearborn, Michigan

As part of its comprehensive Cardiovascular Risk Intervention Program, Ford includes weight control and has developed a program specifically for overweight employees. An interested employee first volunteers to undergo standard screening tests to determine the presence of cardiovascular risk factors such as elevated cholesterol, blood pressure, or obesity. This information is combined with data on the family history of heart disease and on smoking and exercise habits. If these data warrant further attention, the employee is invited to participate in a risk intervention group. Individual consultation is also available. Only mild to moderately overweight employees are treated. Anyone needing to lose more than 30 pounds is referred to outside resources. The program does not have a full-time nutritionist *(39)*.

General Foods, White Plains, New York

A pilot program was tried in 1977 with 12 overweight employees (6 men and 6 women) of General Foods. All signed a contract, putting up $100, which either would be returned to them if they attended classes faithfully or would be sent to their favorite charity if they failed to attend. At 19 meetings, both individual and group sessions, such topics as good nutrition, exercise, risk factors, and behavior modification techniques were presented. At posttreatment, the 6 women had lost an average of 15 pounds and the 6 men an average of 18 pounds. A follow-up six months later revealed that 9 of the 12 had regained their losses. Those who kept the weight off were in an exercise program. Because of the great investment of time in these 12 employees, the program was not cost-effective.

More recently, General Foods offered during the lunch hour a series of five half-hour lectures on weight management. A physician, a nutritionist, and the company's associate manager of health fitness talked about behavior modification and risk factors, diet choices, energy expenditure, and fad diets, and a film on weight management was presented. A total of 90 employees—17 men and 73 women—attended the first lecture. Attendance dropped at subsequent lectures but all were reasonably well attended. The series will be repeated at another plant *(40)*.

Kimberly-Clark Corporation, Neenah, Wisconsin

Weight control is one part of the Health Management Program of the Kimberly-Clark Corporation. Employees at the corporate offices and facilities near Neenah are invited to get a company-paid evaluation of their health risks, using an extensive 40-page medical history; laboratory tests including hemoglobin, blood sugar, cholesterol, triglycerides, liver function, urinalysis, chest X-ray, breathing, skinfold thickness, body density, electrocardiogram, hearing, vision, blood pressure, and temperature; complete physical examination; and treadmill test. They then receive individualized health prescriptions that might include counseling and seminars on obesity. The company has spent $2.5 million to build an office for testing and a large physical fitness facility that includes a swimming pool, 100-meter track, exercise equipment, sauna, whirlpool, showers, and lockers.

The screening is offered annually to all employees. Physical examinations are given every two years for employees 40 and older and every three years for those under 40. More than 60 percent of Kimberly-Clark's 2,100 employees in Neenah have participated in the program *(41)*.

Division of Federal Employee Occupational Health, Public Health Service

The Division of Federal Employee Occupational Health recently held a three-day training program on "Nutrition and Weight Reduction" to provide staff nurses with enough skills and knowledge to plan and implement weight reduction programs in the occupational health setting. Each class was limited to 12 participants.

Topics covered during three sessions were principles of nutrition, psychological aspects of obesity, application of behavior modification to weight control, small-group dynamics, the principles of learning, and the planning and administration of a weight reduction program in the occupational health setting. A nutritionist from the American Heart Association lectured on principles of nutrition and discussed the harmful effects of fad diets and the calculation of caloric requirements. A psychiatric social worker spoke about behavior modification and the techniques she employs in her private practice. Division staff—the health education assistant, Health Education Branch, and the nurse trainer and deputy chief, Clinical Services Branch, covered the other topics.

This training program has been held three times for a total of 34 participants. Nine nurses in the first two programs have started weight reduction groups. These groups range in size from 3 to 10 employees who meet weekly for 8 to 12 one-hour sessions.

Following formal treatment, maintenance meetings are usually held every two to four weeks. Weight losses generally have been one-half to two pounds a week. The program is currently being evaluated by the Division's staff, who will decide whether to expand it eventually to all 148 health units nationwide, for federal employees *(42)*.

U.S. Air Force Hospital, Tinker Air Force Base, Oklahoma

An unusual nonvoluntary behavioral treatment program is being conducted at Tinker Air Force Base for personnel who meet the Air Force's criteria of obesity. Active-duty personnel who have been identified at squadron level as being obese are referred to the treatment clinic at the hospital. Paticipants attend weekly classes, and progress reports are sent monthly to each squadron's weight control monitor. Participants are dropped from the program when they achieve their goals. Maintenance sessions are offered to those who wish to attend. Average weight losses for these groups have ranged from 6 to 10 pounds, depending on the length of treatment (one to four months).

According to its organizers, the program has met with enthusiastic endorsement by unit commanders, as well as the patients themselves, and has lent itself to easy adaptation to the military population *(43)*.

Union-sponsored Weight Loss Programs

Dr. Albert J. Stunkard of the Department of Psychiatry, University of Pennsylvania School of Medicine, is carrying out a weight control program with members of the United Storeworkers Union. They work at Gimbel's 34th Street store in New York City, next door to the union's headquarters. Interested overweight union members have been randomly assigned to various treatment or control groups. Depending on their group, employees may receive weekly or daily behavioral treatment from professional or union leaders at the headquarters or in a traditional medical setting. This study is the first rigorous experimental investigation of behavioral weight control programs at the worksite. Its results may indicate the potential impact that the collaboration of behavioral scientists with union leaders can have on the health of the members (*44*).

Gold King Program

Gold King, an oil production and drilling company in Houston, Texas, requested that the Medical Hospital Weight Control Clinic conduct two series of weight control classes for its employees. The company paid $100 for each of the 22 overweight employees participating. Both series were conducted by a dietitian who was familiar with behavior modification techniques and nutrition information. The company provided a conference room and equipment for showing slides. Employees were allowed time from work to attend eight one-hour classes. The lesson plans for the eight classes follow.

GOLD KING CLASSES

1. Each participant was measured—weight, height, triceps skinfold, and a side-profile photograph. A diet history and behavioral questionnaire were completed. Each employee signed an informed-consent statement. The lecture consisted of an introduction to behavior modification techniques, including self-monitoring, stimulus control, and contingency management. The HELP Your Heart Eating Plan for Weight Control was also introduced. It is a well-balanced eating plan, adequate in vitamins and minerals and low in cholesterol and saturated fat, designed to lower cholesterol and triglycerides by about 10 percent in normal persons. Use of the behavior modification techniques and following the eating plan could result in a weight loss of 1 to 2 lbs per week. It was recommended that participants purchase food scales, measuring cups, and spoons for weighing and measuring food.

2. "Control of Eating Environment" by confining eating to one place in the home, making eating a singular activity, and leaving a little food on the plate at each meal during the next week as a means of demonstrating self-control was discussed in detail. The nutrition lesson consisted of a discussion of the food groups along with servings allowed from each group. The dietitian also designated a caloric level for each participant according to the food intake and diet history each person had filled out the previous week.

3. During "Control When Eating Away from Home," participants were

taught how to choose appropriate restaurants and given guidelines for following the HELP Your Heart Eating Plan. They role-played some ticklish situations, such as what to do when a friend serves a rich dish not on the diet, or when the foods served at a banquet do not fit with the new eating plan. The nutrition lesson consisted of selecting the best choices on the menus of various local restaurants.

4. For the behavior modification lesson, "Control and Change of Food Consumption," participants were encouraged to serve their allowed portion only once, use a slightly smaller plate, and slow the rate of eating so that the meal lasts at least 20 minutes. Formal weight control contracts were distributed. The nutrition lesson was devoted to dairy products; participants tasted various low-fat cheeses available in local supermarkets.
5. The behavior modification lesson was a continuation of stimulus control techniques, "Control of Eating Cues." These included storing food out of sight and not snacking while preparing meals. Jacobson's progressive relaxation technique was also introduced, starting with the relaxation of muscles in the hands, arms, neck, shoulders, and head. This technique was added to aid those who overeat because of stress, tension, and anxiety.
6. The behavior modification techniques consisted of a continuation of how to relax including the imagery of relaxing. The nutrition education lesson was devoted to the selection of low-fat meat, fish, and poultry. Participants sampled homemade lean beef sausage.
7. The behavior modification lesson, "Control of Obtaining Food," included making a menu plan and a grocery shopping list, avoiding eating in the car, and rules for disposing of food. Relaxation was reviewed. The nutrition lesson presented information on the bread and cereal group, preparation of foods, portion sizes, and label reading.
8. The behavior modification lesson centered around maintaining ideal body weight; a maintenance diet was also discussed.

Results of these classes for 22 participants are shown in Table 26.1. Within-group differences were statistically significant for both series—$t(8) = 3.69$, $P < .01$ and $t(12) = 5.44$, $P < .01$, respectively.

These classes illustrate one approach that a concerned company developed for its employees. Continued contact through periodic booster classes may be necessary to ensure continuing losses to reach ideal body weights. Follow-up data will help to determine whether such interventions are lasting.

Although small in scope, such programs may have a decided impact on employee morale and health. The cost-effectiveness of such approaches, however, must await the collection of long-term results.

TABLE 26.1 Results of an eight-week behavioral treatment program at a worksite

GROUP AND PARTICIPANT NOS.	SEX	AGE (YEARS)	HEIGHT (INCHES)	WEIGHT IN POUNDS		
				PRE-TREATMENT	POST-TREATMENT	CHANGE
Group 1 (N=9):						
1	F	24	65	204	198	− 6
2	F	49	64	126	115	−11
3	F	23	66	135	123	−12
4	F	28	64	136	138	+ 2
5	F	26	64	120	116	− 4
6	M	42	67	172	163	− 9
7	F	26	64	154	154	0
8	M	44	71	190	177	−13
9	F	57	64	131	126	− 5
Mean		35.4		152	146	[1]− 6
S.D.		12.7		30.2	29.3	5.1
Group 2 (N=13):						
10	M	52	72	207	202	− 5
11	M	38	70	182	169	−13
12	F	23	62	134	125	− 9
13	M	35	69	207	190	−17
14	M	43	73	212	202	−10
15	F	28	65	115	112	− 3
16	F	29	64	126	124	− 2
17	F	32	62	104	97	− 7
18	F	31	68	163	158	− 5
19	F	30	64	127	125	− 2
20	M	36	73	268	266	− 2
21	M	47	69	189	183	− 6
22	M	37	72	174	164	−10
Mean		35.5		170	163	− 7*
S.D.		8.1		47.6	46.7	4.6

*$P < .01$.

NUTRITION EDUCATION PROGRAMS IN INDUSTRY

Nutrition education programs in industry are rarer than weight control programs. Boeing, Land O'Lakes, and Campbell Soup Company took three different approaches to educate their employees in the principles of good nutrition.

Boeing Company, Seattle, Washington

The Boeing Company has contracted with the ARA Food Service Company to provide a health and nutrition program for its employees. The program has three objectives:

1. Serving foods low in fat, calories, and cholesterol as an alternative to the regular food available at each Boeing food service facility;
2. Supplying vending machines with nutritious snacks such as raisins, fresh fruit, nuts, and seeds, and;
3. Providing nutrition education articles in the Boeing newspaper.

The program is identified in the 85 cafeteria and food service areas by a three- by five-foot rainbow sign with the slogan, "Eat Better and Feel Better." Although the program was initiated only in April 1978, it accounts for 10 to 12 percent of the per capita food sales (approximately 45,000 people eat at the company's food service facilities daily) *(45)*.

Land O'Lakes Company, Minneapolis, Minnesota

Land O'Lakes Company has developed a program using computers to create awareness and interest in diet and exercise. Periodically, computer terminals are made available to employees so that they can talk to the computers. Interested persons keep a three-day record of foods eaten and energy expended. The computer responds with feedback on the employee's food intake and exercise behavior. Each employee is also asked about his weight and activities and given information by the computer about his risk of heart disease. This program has not yet been evaluated *(46)*.

Campbell Soup Company, Camden, New Jersey

The Campbell Soup Company has conducted a large-scale study of the prevalence patterns of lipoprotein and serum lipid abnormalities, the relationship of these lipid factors to the development of atherosclerosis, and the effects of changing some of these factors. Employees are given a blood test, and a family history and other measurements are recorded. About 35 percent of the employees screened have elevated serum lipids. Employees are offered information about a prudent diet—low in fat, cholesterol, and sugar—along with pamphlets on good nutrition to help ensure an adequate intake of vitamins and minerals. Results suggest that about 50 percent of those receiving dietary information are following it after one year, and about 10 percent follow it adequately (47).

EVALUATION OF INDUSTRY PROGRAMS

To date, it has not been possible to evaluate the impact of the few programs being conducted at worksites. As more programs are started, however, it will become important to design them to permit evaluation. Weight control and nutrition programs need to take into account five major factors.

1. *Cost-effectiveness.* With the emphasis on controlling costs increasing, it is important that all costs—including personnel used, testing, time off from work to attend classes, and so forth—be compared with potential benefits, such as reduced disability, absenteeism, hospital and insurance premiums, and turnover; improved employee morale; and direct benefits of pounds lost or cholesterol lowered.

2. *Attrition.* To evaluate a program properly, data on employees who drop out are needed.

3. *Follow-up.* Results of treatment are rarely reported beyond one year after a program is completed. Long-range data are especially important to determine if continued weight loss occurs or if goal weight is maintained.

4. *Confounding variables.* Because there may be uncontrolled or poorly controlled variables, it is especially difficult to attribute changes in employees' weight or diet to the techniques employed. Expectations of employees and group leaders, demand characteristics, and other nonspecific variables make interpretation of results vulnerable to error. It is important to design programs carefully in order to avoid as many problems as possible in the work setting.

5. *Analysis of results.* Weight control and nutrition education programs need to report pretreatment, posttreatment, and follow-up data on each participant. Reporting only average results often obscures true data and may distort the interpretation.

RECOMMENDATIONS

A number of diets low in calories, saturated fat, and cholesterol are available for use in weight control and nutrition education programs. An example of such a diet is the HELP Your Heart Eating Plan *(48)*. The calorie level can be adjusted to allow a person to lose one to two pounds per week or maintain present body weight. In this diet, 40 percent of the calories are in carbohydrates, 20 percent in proteins, and 35 percent in fats, with 10 percent of the fats polyunsaturated, 10 percent saturated, and the remainder monounsaturated. It contains about 300 mg of cholesterol daily. Table 26.2 gives the details of this eating plan.

Basic research on the treatment of obesity and the development of good dietary habits has also been carried out, primarily in colleges, universities, and health clinics. Results of these studies suggest that behavior modification techniques are superior to other forms of treatment in helping mildly to moderately overweight persons adhere to dietary regimens and thereby achieve weight losses. These losses, although modest, are maintained for at least one year after the formal treatment ends *(38)*.

The potential for applying these programs at worksites within industry, government, and the armed forces is enormous. The numbers to be reached are almost unlimited, and with careful planning, the programs could conceivably have powerful effects on the health of participants.

Interest today is shifting from curative to preventive programs, and the worksite offers a unique opportunity to provide both types of interventions. Periodic employee health screenings can identify those who are at highest risk for cardiovascular diseases, and programs might be made available to aid them. Those with elevated cholesterol, for example, might, under a physician's supervision, be offered classes in proper diet and the behavioral techniques for adhering to the diet. Those who need help to lose weight might be offerd similar intervention.

Essentially all dieters could use the same well-balanced, low-fat, low-cholesterol diet, adjusting the caloric levels to individual needs.

In addition to these "at risk" populations, companies might offer interested employees similarly structured preventive programs that could take the form of special lectures, classes, posters or articles in company papers designed to teach good nutrition and to motivate employees to follow healthy eating plans. Foods low in calories, cholesterol, and saturated fat might be made available in a company's food service facilities and vending machines. These foods could be clearly identified.

We studied the effects of offering choices of low-cholesterol dishes in a local Houston restaurant, with a special menu identifying these foods. The calorie content of each low-cholesterol item was printed next to it. Over the year that we measured the sales of these foods, we found that the menu

TABLE 26.2 The HELP Your Heart Eating Plans

FOOD GROUP	REGULAR PLAN	LOW-CALORIE PLAN
Vegetable oil	2 tablespoons daily safflower, corn, soybean, or cottonseed oil	1 tablespoon daily safflower oil
Soft tub margarine	2 tablespoons daily safflower, corn, or sunflower oil margarine	2 teaspoons daily safflower, corn, or sunflower oil margarine
Meat, fish, poultry	7 ounces daily	6–7 ounces daily
Skim dairy products (less than 1 percent butterfat)	unlimited, unless controlling calories	2 servings dairy products
Lowfat dairy products (1–2 percent butterfat)	2 servings daily	none recommended
Grain products and starchy vegetables	4 or more servings daily	4 servings daily
Vegetables, nonstarchy	unlimited	unlimited
Fruits	4 or more servings daily	4 servings daily

MEAT, FISH, POULTRY, EGGS
The leanest cuts of meat are the lowest in saturated fat and calories and are allowed on the HELP Your Heart Eating Plan. Fish and poultry are especially low in fat and should be selected more frequently than beef, pork, and lamb. Egg yolks are the most concentrated source of cholesterol and are limited to two per week. Egg substitutes and egg whites can be used freely since they contain no cholesterol.

DAIRY PRODUCTS
Lowfat dairy products (those containing no more than 2 percent butterfat) and skim milk products are allowed. For persons who need to lose weight, the skim milk dairy products are lower in saturated fat and calories than the low-fat varieties. Cheese, a popular food, is difficult for people to give up; however, if carefully selected, cheese can be included in the low-cholesterol diet. Varieties made with skim milk or polyunsaturated fat are available on the retail market.

BREAD AND CEREAL PRODUCTS
Foods in this group are often termed "fattening," but this label is not really warranted. It is the spread (honey, preserves, gravy, or butter) that doubles the calories and needs to be limited.

FRUITS AND VEGETABLES
Four or more servings daily are recommended from this group, with at least one that is a good source of vitamin C and one a good source of vitamin A.

FATS
Fat is the most concentrated source of calories and should be selected carefully. This food group affects the level of cholesterol in the blood. Large amounts of saturated fat tend to increase plasma cholesterol, polyunsaturated fat helps reduce it, and monosaturated fat has no effect on it.

appealed to a small but consistent number of customers. The restaurant administrators reported that the cost of printing the special menu was the only major expense incurred in this experiment. No additional manpower or time was required, and the ingredients of the dishes were among those routinely stocked. The restaurateurs were pleased with the menu and plan to offer it indefinitely (49).

As nutrition education and weight loss programs are implemented in industries, life and health insurance companies might help collect information about their effectiveness as well as help in the treatment by offering decreased insurance payments as incentives to encourage companies to design and offer programs and to encourage employees to participate in them.

CONCLUSIONS

The potential for developing health promotion programs at the worksite has not yet been tapped. Nutrition education programs for weight reduction and attuned to cardiovascular risk factors related to diet offer particular promise because they can be put into practice so easily in company cafeterias. The combination of a prudent eating plan and the behavioral techniques that teach people how to adhere to such a diet have been developed and can be put into practice. As programs are developed, we hope that they will be designed so that their cost-effectiveness can be evaluated.

REFERENCES

1. Diet and coronary heart disease. A joint policy statement of the American Medical Association Council on Foods and Nutrition and the Food and Nutrition Board of the National Academy of Sciences, National Research Council. *Journal of the American Medical Association* 22:1647–1652 (1972).
2. Inter-Society Commission for Heart Disease Resource. Primary prevention of atherosclerotic diseases. *Circulation* 42:A55–A95 (1970).
3. Select Committee on Nutrition and Human Needs. U.S. Senate, *Dietary goals for the United States.* 2nd ed. Washington, D.C.: Government Printing Office, 1977, p. 4.
4. Stunkard, A., and McLaren-Hume, M. The results of treatment for obesity. *Archives of Internal Medicine* 103:79–85 (1959).
5. Bellack, A. S.; Rozensky, R.; and Schwartz, J. A. A comparison of two forms of self-monitoring in a behavioral weight reduction program. *Behavior Therapy* 5:523–530 (1974).

6. Green, L. Temporal and stimulus factors in self-monitoring by obese persons. *Behavior Therapy* 9:328–341 (1978).
7. Mahoney, M. J. Self-reward and self-monitoring techniques for weight control. *Behavior Therapy* 5:48–57 (1974).
8. Mahoney, M. J.; Moura, N. G.; and Wade, T. C. Relative efficacy of self-reward, self-punishment, and self-monitoring techniques for weight loss. *Journal of Consulting and Clinical Psychology* 40:404–407 (1973).
9. Romanczyk, R. G. Self-monitoring in the treatment of obesity: Parameters of reactivity. *Behavior Therapy* 5:531–540 (1974).
10. Stollak, G. E. Weight loss obtained under different experimental procedures. *Psychotherapy* 4:61–64 (1967).
11. Kazdin, A. E. Self-monitoring and behavior change. In *Self-control: Power to the person* edited by M. J. Mahoney and C. E. Thoresen. Monterey, Calif.: Brooks/Cole Publishing Co., 1974.
12. Nelson, R. O. Methodological issues in assessment via self-monitoring. In *Behavioral assessment: New directions in clinical psychology* edited by J. D. Cone and R. P. Hawkins. New York: Brunner/Mazel, 1977.
13. Richards, C. S. Assessment and behavior modification via self-monitoring: An overview and bibliography. Abstracted in the JSAS Catalog of Selected Documents in Psychology 7:15 (1977).
14. Thoresen, C. E., and Mahoney, M. J. *Behavioral self-control*. New York: Holt, Rinehart, and Winston, 1974.
15. Abramson, E. E. A review of behavioral approaches to weight control. *Behavior Research and Therapy* 11:547–556 (1973).
16. Abramson, E. E. Behavioral approaches to weight control: An updated review. *Behaviour Research and Therapy* 15:355–363 (1977).
17. Bellack, A. S. Behavior therapy for weight reduction. *Addictive Behaviors* 1:73–82 (1975).
18. J. P. Foreyt, ed. *Behavioral treatments of obesity*. New York: Pergamon Press, 1977.
19. Franks, C. M., and Wilson, G. T., eds. *Annual review of behavior therapy: Theory and practice*, vol. 3. New York: Brunner/Mazel, 1975.
20. Hall, S. M., and Hall, R. G. Outcome and methodological considerations in behavioral treatment of obesity. *Behavior Therapy* 5:352–364 (1974).
21. Leon, G. R. Current directions in the treatment of obesity. *Psychological Bulletin* 83:557–578 (1976).
22. Stuart, R. B. Behavioral control of overeating: A status report. In *Obesity in perspective*, edited by G. Bray. DHEW Publication No. (NIH) 75-708. Washington, D.C.: Government Printing Office, 1975.
23. Stuart, R. B., and Davis, B. *Slim chance in a fat world: Behavioral control of obesity*. Professional ed. Champaign, Ill.: Research Press, 1972.
24. Stunkard, A. J. New therapies for the eating disorders. *Archives of General Psychiatry* 26:391–398 (1972).
25. Stunkard, A. J., and Mahoney, M. J. Behavioral treatment of the eating disorders. In *Handbook of behavior modification and behavior therapy*, edited by H. Leitenberg. Englewood Cliffs, N.J.: Prentice-Hall, 1976.
26. Walen, S.; Hauserman, N. M.; and Lavin, P. J. *Clinical guide to behavior therapy*. Baltimore: Williams and Wilkins Co., 1977.

27. Wilson, G. T. Methodological considerations in treatment outcome research on obesity. *Journal of Consulting and Clinical Psychology* 46:687–702 (1978).
28. Hagen, R. L. Group therapy versus bibliotherapy in weight reduction. *Behavior Therapy* 5:222–234 (1974).
29. Hall, S. M. Behavioral treatment of obesity: A two-year follow-up. *Behaviour Research and Therapy* 11:647–648 (1973).
30. Harris, M. B. Self-directed program for weight control: A pilot study. *Journal of Abnormal Psychology* 74:263–270 (1969).
31. Harris, M. B., and Bruner, C. G. A comparison of a self-control and a contract procedure for weight control. *Behaviour Research and Therapy* 9:347–354 (1971).
32. Harris, M. B., and Hallbauer, E. S. Self-directed weight control through eating and exercise. *Behaviour Research and Therapy* 11:523–529 (1973).
33. Jeffrey, D. B. A comparison of the effects of external control and self-control on the modification and maintenance of weight. *Journal of Abnormal Psychology* 83:404–410 (1974).
34. Levitz, L. S., and Stunkard, A. J. A therapeutic coalition for obesity: behavior modification and patient self-help. *American Journal of Psychiatry* 131:423–427 (1974).
35. Romanczyk, R. G., et al. Behavioral techniques in the treatment of obesity: A comparative analysis. *Behaviour Research and Therapy* 11:629–640 (1973).
36. Wollersheim, J. P. Effectiveness of group therapy based upon learning principles in the treatment of overweight women. *Journal of Abnormal Psychology* 76:462–474 (1970).
37. Currey, H., Behavioral treatment of obesity: limitations and results with the chronically obese. *Journal of the American Medical Association* 237:2829–2831 (1977).
38. Jeffrey, R. W.; Wing, R. D.; and Stunkard, A. J. Behavioral treatment of obesity: The state of the art 1976. *Behavior Therapy* 9:189–199 (1978).
39. Ware, B. C., Corporate Coordinator, Health Education Programs, Ford Motor Company. Personal communication, October 30, 1978.
40. Fein, M., Health Fitness Coordinator, General Foods. Personal communication, November 9, 1978.
41. Dedmon, R. E., M.D., Staff Vice President, Medical Affairs, Kimberly-Clark. Personal communication, November 7, 1978.
42. Ederma, A. B., M.D., Director, Division of Federal Employee Occupational Health, Health Services Administration, Public Health Service. Personal communication, October 26, 1978.
43. Reppart, J. T., and Shaw, C. G. A conceptual and statistical evaluation of a new obesity treatment program in a military population. *Military Medicine* 143:619–623 (1978).
44. Stunkard, Albert J., M.D., Department of Psychiatry, University of Pennsylvania School of Medicine. Personal communication, June 2, 1978.
45. McKinley, M. C., registered dietician with ARA Food Service Company. Personal communication, October 31, 1978.
46. Stansfield, A., Consumer Affairs Director, Land O'Lakes. Personal communication, November 30, 1978.

47. Cox, M., and Wear, R. F., Jr. Campbell Soup's program to prevent atherosclerosis. *American Journal of Nursing* 72:253–259 (1972).
48. Gotto, A. M., et al. *HELP your heart eating plan* (booklet). Houston: Baylor College of Medicine, 1975.
49. Scott, L. W., et al. A low cholesterol menu in a steak restaurant. *Journal of the American Dietetic Association* 74:54–56 (1979).

27

Control of Alcohol and Drug Abuse in Industry: A Literature Review

Robert L. DuPont, M.D., and Michele M. Basen, M.P.A.

The increased attention to occupational health in recent years stems from industry's concerns for the well-being of employees and for the maintenance of productivity. These concerns are part of industry's heightened awareness of social responsibility; for example, its involvement in urban renewal, equal opportunity hiring, and programs for the disadvantaged *(1)*. Industry is not only adapting to the changing values and attitudes toward the job and the work world, primarily by young jobseekers *(2, 3)*; it is also concerned with workers' well-being away from the workplace.

The focus in this report is on a broad range of problems associated directly and indirectly with workers' use of alcohol and other drugs that are not prescribed by a physician. Some drugs, such as marijuana and heroin, are illegal; others, such as amphetamines and tranquilizers, are sometimes prescribed. In the range of nomedical use of psychoactive drugs ("alcohol and drug abuse" or "substance abuse"), alcohol-related problems are now pre-

---

This article was first published in *Public Health Reports* (March–April 1980), pages 137–148, and appears here in slightly revised form.

Dr. DuPont is President of the Institute for Behavior and Health, Inc., Bethesda, Maryland. Ms. Basen is a public health analyst with the National Institute on Drug Abuse.

eminent; but other substance abuse is being seen with increasing frequency, particularly among younger workers.

REASONS FOR SUBSTANCE ABUSE PROGRAMS

A review of 17 large occupational drug abuse programs *(4)* revealed that many were initiated by union or management officials who had personally experienced alcohol or drug abuse problems—many were recovered alcoholics, or they had adolescent children who were experimenting with drugs or other related experiences. But the human relations ideology in industry is not purely altruistic. Some employers initiate programs in order to avoid the trouble and expense of grievances, hearings, and arbitrations sometimes spawned by alcoholics and other drug users *(5, 6)*; evidence in this regard has been reported *(7)*. The issue of the potential "social control" aspect of occupational substance abuse programming is explored later in this paper.

The cost of an alcohol or drug abuse problem to the employer is often said to outweigh the cost of operating a program to correct the problem. This claim is seldom documented. Although there are many reasons for inadequate evidence of program effectiveness (as discussed later), the efficiency claim is often used to justify the promotion of a program. Some recent articles cast doubt on the importance of the cost issue. Roman *(8)* concludes the cost-effectiveness arguments may not be as persuasive to industry as previously thought, based on his study of the reasons that executives of large organizations gave for adoption of or resistance to occupational substance abuse programs. Roman also concludes that industry has the wherewithal to conduct sophisticated studies, but such studies have not been done. Clearly, efficiency considerations are not the sole contributors to a decision regarding the need for an occupational program.

The following comment by Dr. R. J. Hilker, medical director for Illinois Bell, encompasses all three reasons in support of alcohol and drug abuse programs:

> With the expansion of benefit and insurance programs, the view that rehabilitation is the responsibility of the community, with the individual employee utilizing community resources to accomplish his or her own rehabilitation, is a shortsighted one. Problems which are common to the community are common to any employee body. If industry takes the position that rehabilitation is not its responsibility, and these employees are simply dismissed, then inefficient, impaired persons will continue to be taken into employment, trained, disciplined and dismissed. The company, meantime, will suffer from absenteeism, inferior service or productivity, management frustration, poor morale and increased insurance costs. Industry, therefore, has at least a business reason to try to reha-

> bilitate employees who have behavior disorders. These disorders often are not detected or are poorly treated by the general medical community. Many studies have shown that behavior disorders can be handled within an industrial setting more efficiently. With early discovery and intervention the illness may be prevented from thoroughly disabling the employee. (9)

EXTENT OF THE PROBLEM IN INDUSTRY

What constitutes a drug or alcohol problem in industry? Contrary to popular belief, drug usage in the workplace is relatively widespread, and it is not confined to blue-collar minority groups *(10)*. Clearly, a definition is needed. The problem is not that a person uses a drug off or on the job, nor even the type of drug; rather, it is the behavior that may be induced by the drug. More specifically, "it is the behavior relative to the performance of [the] job" *(11)*. An employee whose drug or alcohol usage impairs his or her health and interferes with his or her work performance has a problem.

The preceding definition is useful, but not complete. The focus is the job performance, with a clear rationale for the employer's concern and the potential for the development of an organizational response. Numerous related elements of a drug or alcohol problem, including acute drug intoxication on the job and the buying and selling of illicit drugs at the workplace, are important. However, an employee's deteriorating job performance resulting from an underlying substance dependency represents the greatest risk because it is the most prevalent element of a drug or alcohol problem; it is also the most difficult element to measure *(11)*.

Another element of the definition transcends the impact of substance use on job performance and includes more subjective and human issues. These issues encompass the employees' concerns about their own use, or the use by family members, of alcohol or other drugs. Although the primary targets of occupational treatment programs may be those workers whose job performance is impaired by substance abuse and who are therefore identified by their supervisors or co-workers, the programs should also be open to workers who wish to enroll voluntarily, as well as to their spouses and children.

Despite many difficulties—the worst being that data are limited to self-reports, which probably results in an undercount of alcohol and drug problems—several attempts have been made to document the extent of these problems. A 1975 report of the Florida Department of Health and Rehabilitation Services *(12)* stated that about 10 percent of the state's work force were persons whose job performance had deteriorated—half were in the early stages of alcoholism or overtly addicted to alcohol, and the remainder were experiencing behavioral or medical disorders (including abuse of other drugs). Another 1975 report, by Booz, Allen and Hamilton *(13)*, stated that of the 76 million people in the U.S. work force, 3 to 7.6 million suffered from alcoholism.

Some researchers have estimated the extent of alcoholism and drug abuse in various kinds of work organizations. Cahalan and Cisin *(14)*, in a survey of Navy personnel, found that 19 percent of enlisted men and 9 percent of enlisted women had experienced either critical or very serious consequences from alcohol consumption during the three years before the study. Hitz *(15)* concludes that "some occupations seem to provide acceptance or encouragement of drinking patterns and problems which may not be encouraged or accepted elsewhere," and that drinking problems were far more common among "lower blue-collar workers"; this finding was confirmed by other researchers *(16, 17)*. Roman's study of a national sample of more than 500 executives in large, private businesses *(8)* discloses that just under 10 percent of them believed that the prevalence of alcohol-related problems in their organizations was as high as 5 percent of their employees—25 percent thought it was lower than 1 percent.

The 1971 New York State Narcotic Addiction Control Commission study *(18)* is widely quoted. This study, which excluded alcohol, found that marijuana was the drug most often used, followed by minor tranquilizers and barbiturates; the issue of job impairment caused by abuse of such drugs was not specifically addressed. However, Trice *(19)* constructed a definition of abuse, using the drug prevalence findings of the commission's study and his knowledge of the impact of various drugs, to obtain a prevalence estimate of drug abuse. From the available data on use and effect on behavior of heroin, barbiturates, and other drugs, Trice concluded that about 1 to 2 percent of the workers in New York City in 1971 were impaired by drug abuse other than alcohol.

In a 1974 study of a national sample of 197 firms that focused on management's perceptions of drug use *(20)*, both management and employees reported an awareness of work problems related to use of marijuana (65 percent), amphetamines (39 percent), and barbiturates (35 percent). Further, in a sample of employees asked to report their own drug use, almost 75 percent stated that they were currently using an illegal or nonprescribed drug. Unfortunately, this study also concentrated on patterns of use and did not identify problems associated with this use.

Steele *(21)* found that although the literature claimed extensive drug usage and a major drug problem in industry, a comparison of the claims with the results of various surveys did not conclusively support this assertion. However, he noted that the results of early regional surveys in metropolitan areas seemed to be in concurrence *(22)*; yet, in other surveys the persons interviewed distinguished between moderate usage in their companies and drug problems in industry as a whole *(23–25)*. Other researchers have reported that relatively few officials perceived a drug problem in their organizations *(20, 26, 27)*.

In a recent study of American young men, it was found that those employed were somewhat less likely to have used drugs (other than alcohol) nonmedically than those who were unemployed *(28)*. Even if we accept this finding, recent general population surveys suggest that nonmedical drug use is now far more frequent than many people realize. For example, about 30 percent of all 18- to 25-year olds had used marijuana at least once within the past month. In 1977 *(29)*, 11 percent of the nation's high school seniors used marijuana every day, as opposed to about 6 percent who used alcohol every day. While marijuana use far exceeds the frequency of use of any other nonmedical drug (except alcohol and tobacco), the use of tranquilizers, stimulants, and depressants is no longer uncommon. For example, more than 18 percent of Americans between the ages of 18 and 25 reported having used sedatives nonmedically, and nearly 3 percent of those over age 26 reported similar use of drugs. Recent surveys of the general population have found a dramatic increase in the levels of nonmedical drug use within the past decade, primarily in the under-20 age group. Although the tradition of higher use rates for men, minority group members, the poor, the young, and urban dwellers still exists, all the gaps are now narrowing; usage among the previously lower-use segments of the population is increasing most rapidly *(30)*.

Caplovitz *(31)*, in an intensive study of working addicts in treatment, found that the characteristics of working addicts were more similar to those of other workers than to those of nonworking addicts. For example, they tended to be older, better educated, more often married, and more likely to be white than the nonworking addicts in the treatment population. Almost all were addicted to heroin, used high doses of heroin, and were using more than one drug—61 percent were polydrug users, and more than one of every five used at least three illegal drugs in addition to marijuana.

Caplovitz also found that the occupations of his sample of working addicts were fairly similar to those of the general population. Some were employed in a variety of industries, including government, but most were in retailing and manufacturing. Most interesting was that despite their habits, many of these addicts held onto their jobs for some time. Some 68 percent held their jobs for a year or more. However, more than half (53 percent) admitted that their drug habit caused them to lose days at work. Ironically, most addicts (64 percent) believed that their supervisors thought that they were doing a very good job. But 5 percent said they had injured themselves, 4 percent had injured someone else, and 7 percent said they had damaged equipment because of their drug usage. The findings of this and other studies suggest that the stereotype of the heroin addict as a person who is highly unstable and unable to hold a job must be revised.

The impact of drug abuse on work performance varies widely, depending on frequency and amount of use and type and potency of drug (or alcohol). In some instances, individual reactions are also influenced by the setting in which use occurs. In the case of marijuana, it can be assumed that regular use is accompanied by impairment of job performance *(10)*. Heavy, regular use of marijuana produces problems that are related most urgently to driving and other complex psychomotor performance, to studying, and to interpersonal relations *(32)*. According to Chein:

> There is no simple or single effect of opiates on work and productivity. Instead, a variety of behaviors vis-à-vis work may occur when a person is regularly using opiates. Whatever behavior we observe in a particular addict resulted not merely from opiates, but rather as a consequence of interactions between his needs and motives for addiction, his personality structure and the neurophysiological effects of the drugs. *(33)*

In fact, involvement of the heroin addict in the addict subculture—in the interest of maintaining his supply—has negative effects on job performance beyond those produced by the drug itself.

Another issue complicates the question of impact of abuse of drugs on job performance. Devenyi and Wilson *(34)* state that many alcohol abusers are also abusers of barbiturates. The effects of barbiturates, especially if used with alcohol, can have severely detrimental effects on job performance. Trice *(19)*, nevertheless, concludes from available evidence that of all the drugs, abuse of alcohol overshadows the others in terms of impact on job performance. He states that alcohol use, and especially long-term abuse, impaired those cognitive functions required for efficient job performance. In a 1976 study of the influence of alcohol on work performance, Threatt *(35)* examined the effects of alcohol use on various aspects of human behavior. He concluded that long-term alcohol abuse created problems beyond impairment of sensory-motor skills and intellectual performance. The physical deterioration from alcohol addiction is well documented. Physical illness due to alcohol can result in absenteeism and ineffectiveness on the job, while psychological impairment can result in poor decision-making and reduced output. Impaired judgment is associated with higher accident rates, mistakes, and increased workload for other workers. In sum, employee alcohol or drug problems, or both, affect job performance in many ways, including late arrivals and early departures, absenteeism, poor judgment, accidents and safety hazards, erratic and decreased productivity, failure to meet schedules, lowered morale, resentment among other employees, waste of supervisors' time, and damaged customer and public relations.

INDUSTRY'S RESPONSE TO SUBSTANCE ABUSE

Industry has increasingly responded to the problem of alcohol and drug abuse in the form of company policies or programs. "Programs" range from the promulgation of written policies with respect to the organization's response to substance abuse to highly developed, internally staffed programs offering treatment services. Most programs are viewed as part of an employee-employer benefit package designed to identify, motivate, and refer at an early stage those employees with personal-medical problems that contribute to unacceptable patterns of job performance. The assumption is that such programs assist both employers and employees. Employers benefit because they have a control system to identify and offer help to troubled employees, and employees benefit because they are given an acceptable alternative to disciplinary action.

The Third Special Report to the U.S. Congress on Alcohol and Health discusses the following major goals of occupational alcohol programs:

- To reach employed problem drinkers in order to reduce the cost of poor performance and absenteeism associated with their drinking
- To minimize grievances and arbitrations associated with employee alcohol problems
- To recover the health and efficient job performance of valued employees
- To provide assistance to the families of employed problem drinkers (and/or to the family members with drinking problems)
- To intervene early enough to obtain substantial rehabilitation *(36)*

Occupational alcohol programs and policies have been broadened recently to include use of other drugs—primarily, abuse of prescription drugs and, increasingly, polydrug abuse. Such programs are known to have a "broad-brush" approach. They are also often referred to as "troubled employee" or "employee assistance" programs *(37)*. The emphasis of these programs is early identification and intervention in the workplace, thus allowing for possible identification of an employee who is experiencing the early stages of a developing personal problem. The job performance is usually affected early. Therefore the workplace can be viewed as an important location for early detection—and possible prevention—of substance abuse problems. In addition, through this mechanism the employee's family can have access to appropriate substance abuse services.

Four models were identified in a recent classification of occupational alcohol and drug abuse programs: consultation only, assessment-referral, diagnostic-referral, and diagnostic-treatment (inpatient and outpatient or outpatient only) *(38)*. According to Shain *(1)*, current substance abuse program approaches, components, and characteristics are as follows:

Program approaches: Alcoholism only and employee assistance.

Program components: Written policy, labor-management involvement, companywide information and education program, supervisory training, uniform identification and referral procedures, availability of treatment resources, and follow-up procedures.

Program characteristics: Degree of emphasis on "early detection," use of constructive confrontation, location of program in organizational structure, and nature of relationship with treatment facilities.

In brief, the dominant occupational program strategy is as follows. The most essential element of this strategy is the "constructive confrontation" of employees whose job performance has been deteriorating *(17)*. Supervisors are encouraged to present the facts of deteriorating job performance to the employee, with offers of whatever health or counseling services are available, including a description of the alcoholism or drug abuse program. Job impairment is the major focus *(1, 17, 38, 39)*. If job performance continues to deteriorate after referral, the supervisor informs the employee that job penalties will occur, again offers and explains rehabilitative services, explains drug and alcohol abuse policies, and emphasizes that the use of these services is optional.

Trice and Beyer *(17)* report that a large majority of actual policies call for the alcohol and drug abuse program staff to develop referral relationships with community treatment facilities. Referral constitutes the second intervention in the program. Trice and Beyer further pointed out that in unionized companies impaired performance is defined within the framework of collectively bargained contracts and agreements. In short, these programs are based on the assumptions that the most clear-cut mechanism for identifying problems related to alcohol or drug use is the supervisor's awareness of impaired performance; alcoholism or drug use should be regarded as a medical problem; regular disciplinary procedures of poor performance should be suspended while an employee seeks assistance; and return to adequate job performance is the sole criterion for judging successful outcome *(36)*.

According to Trice *(16)*, the use of this general strategy spread slowly during the 1960s. In 1970, more than 100 companies had such policies in operation, and since the early 1970s, the number increased dramatically. However, Trice pointed out that the number of programs

> still totals no more than 300 to 400 among the larger manufacturing companies, banks, utilities, merchandising, transportation, and life insurance companies. If programs in smaller companies, consortia of small firms, and union-initiated programs are added, the total number of well-implemented programs in this country is probably no more than 600. *(16)*

Trice states further that although this was a small proportion of the nearly 500,000 U.S. work organizations that employ 100 or more persons, the increase in job-based programs was substantial. Another source *(36)* estimated that from 1970 to 1973 the number of occupational alcohol programs expanded from 50 to around 500. By mid-1977, the number of organizations with some type of program had increased to nearly 2,400, with 2,000 in the private sector and 400 in the public sector. According to a recent survey *(40)* of a sample of Fortune 500 companies, the proportion of sampled companies reporting having some type of program to identify and help problem drinkers climbed from 25 percent in 1972 to 34 percent in 1974 and to 50 percent in 1976. Although the data indicated that many of these programs needed substantial upgrading, there was also strong evidence of executive involvement and support for them. There were no reports of union resistance to programming efforts.

With regard to occupational drug abuse programs, management attitudes have shifted to a more humanistic perspective from the earlier policy of immediate termination of employment *(20)*. Rush *(25)* found that only 21 percent of 222 companies advocated immediate dismissal, and Johnston *(24)* reported that 23 percent of his sample of 134 employers advocated this policy. Some companies had an informal policy of referring drug users to external rehabilitation sources, but few had formal referral programs. Johnston states that 36 percent of the 134 employers referred users to external treatment sources, and Rush *(25)* reports 35 percent of the employers referred users for rehabilitation. Steele's efforts at gauging union attitudes and commitment reveal that 32 percent of a total sample of 400 respondents had education programs, 46.2 percent had referral policies, and 26.2 percent had union counseling programs for drug users *(21)*.

Trice and Beyer *(17)* consider the Federal Civil Service health program an encouraging example and model for other employers. The program, created by legislative mandate (Public Law 91–616 of 1970 and Public Law 92–255 of 1972), calls for Employee Alcoholism and Drug Abuse Programs for federal civilian employees. A recent breakdown of current active health program efforts for all employees of the Department of Health, Education, and Welfare *(41)* showed that only a limited number are being reached by the health program. The reasons given for limited program growth include inadequate agency resources, lack of visible commitment by top management, geographic dispersion of employees, and lack of coverage for substance abuse by federal health plans.

In 1975, there were 209 substance abuse programs in government agencies and 531 in the private sector *(13)*. The breakdown was manufacturing, 66 percent; transportation and public utilities, 11 percent; business, education, social, and health services, 10 percent; finance, insurance, and real estate, 6 percent; and miscellaneous, 7 percent.

Trice and Beyer *(17)* conclude that the anticipated resistance to alcohol and drug abuse programs among top management and union leaders in the 1960s was exaggerated. Instead, they believe that there was unfamiliarity with and apathy toward occupational programming, rather than outright rejection. Dr. Paul Sherman, president of the Association of Labor Management Administrators and Consultants on Alcoholism (ALMACA) is a bit less optimistic *(42)*. He states that despite a growth of from 300 occupational programs in 1971 to more than 1,500 in 1977, this number represented a very small proportion of the 1½ million U.S. businesses. Moreover, if the programs could be analyzed, perhaps 300 would be found to be operating effectively. Sherman cites stigma and the lack of hard data as major impediments to widespread implementation of substance abuse programs.

PROGRAM IMPACT

What data are available on the actual impact of substance abuse programs? How much do they cost, how many people do they reach, and what are their limitations and successes? There are a number of dimensions of the success of a program. Depending on the motivations behind the establishment of a program and the particular objectives of a program, different groups—employers, employees, unions—judge success in different ways. The essence of a successful occupational program is that it continues over time and is active in rehabilitating its employee participants *(43)*. The extent of implementation and continuous functioning can be observed in many ways, primarily by documentation of the rates and types of casefinding and case disposition and the nature, quality, and rates of desirable outcomes.

The essential question is: "What kind and what amount of intervention works best for what kinds of employees in what kinds of environments?" *(43)*. Specific indicators useful in answering this question are employment status, level of alcohol or drug involvement, job performance level, criminal involvement, disciplinary action, accidents on the job, sick leave and sick benefits, grievances, and unauthorized absences. Indirect indicators are criminal involvement, accidents off the job, marital stability, relationships with children, and levels of psychological and social functioning. As yet, no study has incorporated all of these indicators in an evaluation of an existing program. The evidence about the effectiveness of occupational programs is generally fragmentary *(43)*.

Data collected in a National Institute on Alcohol Abuse and Alcoholism (NIAAA) project from 15 private organizations with alcoholism programs underlined the positive impact of these programs on their clients *(36)*. A second phase of the same study indicated that employers' investment in such programs would result in cost savings. A survey of alcohol and drug abuse programs in the railroad industry indicated that rehabilitation rates averaged 69 percent of referrals. However, the NIAAA report cautions that although many studies show substantial success—an estimated 70 percent of referrals—they rarely mention the unsuccessful 30 percent. Trice *(16)* concludes that despite the lack of compelling evidence, job-based programs do motivate drug or alcohol abusing employees to seek rehabilitation and to remain in the treatment program long enough to secure significant results. Hiker *(44)*, for instance, used a time-series design and compared data on job performance, illness absences, promotions, sobriety, and accidents five years before and after intervention with 402 employees. He reported dramatic before-after differences. Unfortunately, no comparison groups were used. In recent testimony *(45)*, it was reported that general estimates of rehabilitation range from 50 to 60 percent. Bethlehem Steel has reported 60 percent success, and DuPont Corporation reported 66 percent of 950 alcoholics rehabilitated (Florida Department of Health and Rehabilitative Services, 1975). The NIAAA *(36)*, points to the support given to occupational programs by a majority of surveyed executives. However, it is possible that executives in companies with programs tend to claim the success of programs beyond what "evidence" may support, since they have sanctioned these programs.

In sum, the weight of the evidence suggests that occupational programs are relatively effective. And yet, until very recently, the evidence of success for occupational programs was restricted mainly to measurements of job performance, and we do not know how representative the treated population is of the total number of people who could benefit from such programs *(43)*.

It cannot be said with confidence that financial returns to employers who use these programs outweigh their costs, although most programs make these claims and there are few data to refute them. Most research in this area has been unsophisticated. It has been estimated that problem drinking costs industry $1 to $8 billion, and that costs associated with responding to the problem are far less *(46)*.

In several studies the savings resulting from the establishment of an alcohol and drug abuse program were estimated. For example, General Motors' Oldsmobile Division noted a saving of $226,334 as a result of a reduction in lost man-hours *(47)*. Indirect costs are computed with indicators such as improved job performance and reduction of accidents as proof of savings *(42)*. Winslow and associates *(48)* report that suspected problem drinkers were 16 times as costly to insurers than were problem-free employees. Problem drinkers also made a significantly greater number of medical clinic visits, and they were rated lower in percentage of potential by their immediate supervisors.

Program costs related to alcohol and drug abuse treatment for the Federal Civil Service are estimated at $5 per employed person ($15 million) annually with potential cost savings estimated at between $135 and $280 million annually *(49)*. Major American commercial insurance carriers estimate that for every dollar spent in rehabilitation efforts, $5 are ultimately saved *(50)*. Wrich *(37)* estimates that long-term costs over a 25-year period of an employee assistance program with 1,000 employees are $426,740.

As Schlenger and Hayward *(51)* point out, the reliability of any estimate depends on the methodology from which it was derived. Occupational program cost estimates usually are not made statistically. One exception—the evaluation of a military program in terms of costs and benefits *(52)*—has been cited by Roman *(40)* as evidence of promising work.

PROGRAM LIMITATIONS

What are some of the current limitations of occupational programs? What has prevented the more rapid acceptance and development of these programs by a majority of corporations? For one thing, the basic programming model that stresses supervisory confrontation on the basis of deteriorating performance is not appropriate for a number of occupational and professional groups *(41)*, including executives, most professionals, and those who work in isolated settings and small businesses.

Another difficulty is that of determining the role of the unions in program planning, development, and maintenance. Trice *(16)* states that recent research has demonstrated that the simple presence of an interested and involved union is significantly associated with greater use of a drug and alcohol prevention policy by line managers. In addition, where a company is unionized, and line managers know that the union has taken a position in support of the policy, managers are more likely to use an alcohol or drug abuse policy. Unfortunately, a review of many company policies on alcohol and other drugs of abuse showed a relatively low level of union participation *(17)*. However, in many cases labor has been willing to participate in policy and program development *(53)*. Indeed, Trice and Beyer *(17)* cite numerous examples of specific union-initiated policies. There are also an increasing number of union-initiated and union-operated programs for alcohol and drug abusing employees.

But a great deal of sensitivity exists among union members about the employee assistance program model. Because these programs expand an alcoholism policy's coverage to a wide variety of behavior problems, it is feared that management can "control" legitimate forms of dissent. Moreover, union officials view this situation as possibly leading to new and complex collective bargaining and grievance problems. Also, the expansion of such policies and programs could be viewed as an invasion of "turf," since the labor movement has been providing a wide range of services to union members for years. To complicate the issue further, the American labor movement is not one body with one opinion. Many local labor groups set their own policies and form their own programs. In sum, the evidence suggests that while both sides agree with the goals of alcoholism rehabilitation, a number of institutional constraints must be coped with before such programs become more widespread *(54)*.

In a similar vein, there is concern regarding the potentially compulsory nature of substance abuse policies or programs. Because employers can influence the personal behavior of their employees through such policies, the rights and responsibilities of the employee and employer are important issues. Many believe that the right to intervene grows from the employer's right to expect adequate job performance. When drug or alcohol abuse results in impaired performance, the employer has the right to intervene. On the other hand, care must be taken to respect the rights of the employee. Clearly, the employer has the right to intervene only if drug or alcohol use unmistakably impairs job performance. The employer must respect the privacy of the employee.

In a study of the differential use of an alcoholism policy in Federal organizations, by skill level of employees, Trice *(54)* found that the actual use of the policy was greatest in low-skill installations. Although there are many possible explanations for this, Trice concludes that if policies were used as behavior control, a dangerously discriminatory control policy would be in effect.

Another factor in the retarded growth of occupational programs is the limited number of employee health benefit plans that cover alcohol and drug abuse. Hallan and Holder *(55)* report a recent survey of 31 large companies having continuing occupational programs: 30 of these companies made specific provisions for inpatient care, about three-fourths provided benefits for special treatment centers (for example, care in an alcoholism treatment center), but only 15 covered the costs of outpatient care. These authors noted that the mere existence of a benefit structure that could cover the costs of alcoholism effectively in no way assured that such benefits were actually used, nor was there any indication of the extent to which total alcohol treatment costs were being met by insurance benefit payments. The survey findings, however, did indicate that benefit plans are surprisingly liberal. Hallan and Holder conclude that the health insurance industry can respond to occupational program needs by providing broadly based health insurance plans. Unfortunately, the survey findings were severely limited because the data were based on only 31 firms.

Cost is often the reason given for excluding alcoholism treatment in company insurance health plans. There is some evidence from a pilot effort in California, which covered state employees for alcoholism benefits, that for every $1 spent to treat alcoholics an estimated 41 cents was saved in health care for nonalcoholics. If the State of California had not paid all the costs for the program during the experimental two years, the additional average annual premium for each enrolled family would have been only $2.05, or 17 cents a month, to cover the total cost of treatment for alcoholics *(55)*.

As of 1976, 13 states had passed legislation mandating that insurance carriers provide coverage for treatment of alcoholics *(56)*. Also, the Health Maintenance Organization (HMO) Act of 1973 required that all HMOs receiving federal assistance must include alcoholism services in their benefit package. The major national health insurance proposals introduced in the 93d Congress have also included the requirement for appropriate alcohol (and drug) treatment coverage. In organized labor, more than 1¼ million auto industry workers and their families have Blue Cross coverage for the treatment of alcoholism and drug abuse.

Although coverage of drug abuse services, like alcoholism, is improving, a number of problems remain. Some reasons for the traditional lack of interaction between the private health insurance industry and drug abuse treatment programs are: the notion of drug abuse as a disease was highly controversial; insurers questioned the professional status of individual providers of treatment for drug abuse; insurers were uncomfortable with the setting in which most drug abuse treatment services were rendered; and insurers anticipated uncontrollable costs for continued treatment because of the high rate of recidivism *(57)*. The extension by Blue Cross/Blue Shield of Michigan of substance abuse benefits to 1.4 million auto workers is evidence of a more favorable future. Five states have enacted laws encouraging or requiring private insurers to offer drug abuse treatment benefits, and, generally, restrictions that previously limited coverage drastically are being lifted *(57)*. Among the 26 mature HMOs now operating, 16 specifically cover drug abuse services. However, a survey of commercial carriers *(57)*, revealed that they were even less likely than Blue Cross to cover drug abuse services. Of the 174 companies surveyed, 38.5 percent covered drug abuse services in the same manner as other services; 15.5 percent totally excluded drug abuse services, 17.2 percent provided coverage with stringent limitations, and 28.8 percent did not respond to this question.

Perhaps the major factor inhibiting the expansion of occupational programming has been, until just recently, the dearth of research and evaluation efforts. Because most occupational programs are voluntary and are initiated in various kinds of organizations, which have different objectives, it is difficult to define what a program really is and hence, whether or not it is successful. Furthermore, even under the best of conditions, the need for confidentiality often limits access to data and thus precludes the ability to address many crucial research issues *(36)*.

The research and evaluation that has been done was limited by serious methodological problems. The use of "penetration" rates has been a major problem in evaluating the impact of occupational programs. The penetration rate is a measure of the extent to which the program is reaching its target population. The formula for determining this rate either relates the size of the identified problem group in a given industry to the size of the workforce as a whole or relates the identified group to an estimated population at risk within the work force. Unfortunately, prevalence estimates of the total number of employees with drug or alcohol problems in the targeted work force are required for determining penetration rates, and such estimates vary widely. Other factors also enter into determining penetration rates; for instance, the establishment of a program may reduce the number of employees with problems. Thus, formulas for computing penetration rates must be based on the various stages of program development *(51)*.

Defining program success is a second problem in research and evaluation. Success has been defined in various ways, usually as (*a*) significant improvement in job performance by the treated employee or (*b*) modification of drinking or drug-taking behavior. Further serious problems arise because there is a lack of comparability of findings between studies (definitions of problem behavior are not sufficiently specific), criteria for successful rehabilitation are rarely given, clients' characteristics are rarely described, and follow-up intervals vary enormously *(1)*.

Another difficulty is determining the efficiency or costs and savings of occupational programs. According to Schlenger and Hayward *(58)*, reliable cost information is scarce because of the different kinds of records kept by occupational programs, a sensitivity to keeping individual case records, and the difficulty of measuring both the direct and indirect costs of employees' drug or alcohol abuse. These authors also explored the problem of experimental design. The methodologies of most studies involve a before-and-after comparison of persons who have participated in programs. However, any observed changes cannot be attributed positively to the programs because many employees, including those who were not in programs, may have improved over the time studied. Also, the effects of different components of a program are usually not isolated in most studies of program impact.

PROGRAM ISSUES

In view of the experiences of currently operating occupational strategies, what options are open to those who hope to promote substance abuse programs? No one program model is appropriate for every worksite. The development of the model depends on the type of target population and the type of sponsoring organization. The following are essential characteristics that enhance the effectiveness of substance abuse programs *(43)*: written policy; clear procedures; endorsement by top management and union executives; a joint union-management committee; education programs for management and supervisors, union executives and stewards, and employees and families; effective communication at all levels; an active, committed coordinator; informal or formal counselors, or both; active involvement in Alcoholics Anonymous; backup residential treatment service; good liaison with community services; and periodic assessment and updating of the program.

Shain *(1)* advocates that a "model" programs's goal should be to help people achieve and maintain satisfactory health and job performance; it also should focus on the causes of deteriorating health and job performance, with particular emphasis on alcoholism and drug abuse, and adopt the strategy of constructive confrontation. Shain also recommends that organizations establish a second method of case-finding—the voluntary strategy—at the same time they adopt the strategy of constructive confrontation. The voluntary strategy would encourage self-referral into the program, with assurance of confidentiality. Thus, intervention and rehabilitation could take place before the problem affects job performance and health to the extent that constructive confrontation is warranted. The voluntary strategy offers an attractive potential for shifting workplace intervention closer toward primary prevention of alcoholism and drug abuse.

The constructive confrontation policy is, of course, an effective tool for other reasons. This expanded concept of occupational alcohol programs has resulted in identifying other personal problems among the workforce, and it offers the employee appropriate assistance for other difficulties that may affect job performance.

The advantages of union involvement in initiation and operation of programs have been stated. Without active union participation, programs are open to abuse in industries employing large numbers of lower occupational status workers who are easily replaceable *(59)*. The evidence suggests that both sides are in agreement with the goals of alcohol-drug abuse rehabilitation. Therefore, recommended are increased efforts to stimulate joint management-union committees to develop or monitor these programs, or both, and the development of a specific coordinator role for alcohol-drug abuse policy in both the union and management organizational structure *(17, 36)*.

Another important issue in the acceptance and expansion of occupational programs is that of organization and support. If occupational programming is a movement, it is not a highly organized movement with close ties between those who work in it *(59)*. The movement will have to be organized, and an attempt must be made to obtain consensus on future directions in programming. For instance, it has been suggested that occupational program consultants (OPCs) change their focus from marketing and advocacy to voluntarism. In other words, OPCs should be consultants to business leaders to help them establish or improve their programs. Particularly with respect to drug abuse treatment, occupational program concepts must be promoted in both the work and treatment worlds.

Another way to promote these concepts is to create a position for an industrial specialist in the treatment setting. This person would understand the internal company policies and respect their place in the referral and treatment process and thus would be an important link with the work world. Trice *(19)* labels such persons "brokers" and defines their role as providers of reliable, objective information regarding treatment process and outcome to the work world. He specifically suggests giving research grants to students to accomplish this linkage. One way to convince the involved parties of the importance of this linkage is to demonstrate that the extension of services to the employee's family members—a service that can be provided through the work setting—will have a positive impact on the community. Moreover, occupational programs may indirectly improve the quality of treatment service in the community by increasing the available third-party payments.

Trice *(19)* and Sherman *(42)* agree that apathy and unfamiliarity with occupational programs and the problems of the drug and alcohol abusers in general are major reasons for the slow growth of occupational programs. They recommend massive public relations and education programs for labor and management. A DHEW (now DHHS) report similarly suggests massive education in government organizations, based on a study in which it was found that lack of familiarity with the alcohol and drug abuse policy was associated with underutilization of available resources *(36)*. The report further advocates establishing an office of employee assistance programs to demonstrate agency management commitment to the program.

NEEDS OF SPECIAL POPULATIONS

To reach special populations, including women, young drug users, polydrug users, small executive/upper echelon staffs, and persons in small businesses, it is necessary to modify the basic model. About half of the work force is employed in small business. One approach advocated by the Addiction Research Foundation *(43)* and others is the consortium, in which a group of employers or joint employer-union groups in a geographic area establish a collaborative alcohol and drug abuse program. The success of such a consortium depends on two basic components—sharing of both fiscal and governing responsibility.

The great number of women in the work force implies unique alcohol and drug abuse problems. Few women are included in the available data on persons identified and helped by occupational substance abuse programs. Some observers believe that women are primarily in occupational settings that do not tend to have policies and programs, whereas others propose that occupational programs are not as applicable to women as they are to men. At this juncture, it is desirable to tailor the substance abuse model in order to experiment with a variety of approaches for reaching out to women employees.

Polydrug abuse has become a major problem; thus the focus of treatment on a single drug dependency in any context must take this into account. Although polydrug abuse is more common among the young, it also occurs among older age groups. Many of today's youth smoke marijuana morning, noon, and night. This labor force of the future must be of increasing concern to employers; at present it is not targeted by occupational programs. Since the traditional program assumes that an employee's value to the organization is based on substantial training and time investment, this value often does not extend to the youthful abuser. Moreover, young employees may not respond favorably to constructive confrontation—they have less time invested and in some cases a different work ethic. Here too, variations on the traditional substance abuse program model are in order.

FUTURE RESEARCH

The maintenance of occupational programming as a viable movement depends largely on demonstrable success *(59)*. Demonstrating success will be a research problem. Trice and Beyer *(17)* note that most research and evaluation efforts have failed to go beyond time-series patterns. They suggest the use of comparison groups strategies, as well as examination of the impact of different types of treatments on employees referred from the work force and the differential impact of various program components. Trice and Beyer also discuss the problem of quantification of benefits in continuing attempts to determine program efficiency. They point out the deficiency of workplace records. In addition, there is a clear need for reliable data about the use of drugs other than alcohol in the workplace. Although it is reasonable from current evidence to expect alcohol to be the primary problem, there are obvious impacts from the use of other drugs.

Further research also is needed to compare the impact of occupational programs with that of other programs designed to identify and refer the problem drug abuser or alcoholic. And the elements contributing to the positive or negative operations of a program need to be identified and evaluated. Furthermore, the influences of occupational drug and alcohol programs on surrounding communities must be determined.

INSURANCE COVERAGE

Several methods by which occupational programs can positively influence third-party payments have been suggested. One strategy is to convince carriers to voluntarily provide coverage for appropriate and sufficient alcohol and drug abuse services to employers. This strategy is feasible, since a major factor in voluntary coverage will be that of competition between carriers. A second strategy is to encourage major purchasers of health insurance to demand such coverage. This strategy can be most effective because employee representatives are continuously seeking improved fringe benefits. A third strategy is to encourage the enactment of mandated health insurance coverage for alcohol and drug abusing employees by state legislation *(55)*. Obviously, numerous arguments can be raised against inclusion of such employees, but it is recommended that employers, labor union representatives, and representatives of major carriers in target states be invited to meet to become acquainted with the possibilities for and the nature and costs of such coverage.

COMMENTS

Within the past decade, two major social developments have greatly influenced society's responses to alcohol and drug abuse. First, there has been a dramatic increase in public acceptance of alcoholism (and to a much lesser extent, drug abuse) as an "illness" requiring treatment, rather than as a moral failing deserving punishment. This shift, often misunderstood in the medical and scientific communities, is the core concept that has led to the reduction of the stigma against alcohol and drug abusers. This stigma had been a major inhibitor of progress in the substance abuse field in the workplace and elsewhere. Public knowledge that everyone is vulnerable to alcohol and drug problems (rather than just the "bad" or the "weak") has encouraged wider support and relatively rapid growth of substance abuse programs in recent years. The second major development has been the unprecedented increase in the nonmedical use of psychoactive drugs—primarily, but by no means exclusively, among the nation's youth.

These two developments point the way to some new directions in the future. Emphasis on further reducing the stigma and the inclusion of drugs other than alcohol are vital to the success of substance abuse programs. But these broadened concerns will not be easy to accomplish because of the important differences between segments of the substance abuse population—many differences are related to the potent variables of age, sex, race, and social class.

The illegality of nonmedical use of drugs other than alcohol poses serious and largely unresolved problems. Action programs must retain sufficient specificity to meet the needs of nonmedical drug users. For example, it makes little sense to put people who are losing weight, stopping smoking, or quitting heroin together with recovering alcoholics and telling them they all have the same problems of "substance abuse" or "behavioral disorders" or that they are "troubled employees." Thus, although the move toward greater integration is reasonable in terms of management and program techniques, it is often impractical clinically.

Finally, the newly emerging concern for health promotion—also called the "prevention" or the "wellness" movement—offers a promising new opportunity for drug abuse and alcohol programs in the workplace. This new, broader focus provides an escape from many of the problems associated with the earlier preoccupation with the involuntary model of drug and alcohol programs. Turning that coin over, it may also be possible for some of the newer prevention programs to learn some useful lessons from the already functioning drug and alcohol programs in the workplace—including the role of "constructive confrontation."

REFERENCES

1. Shain, M. Occupational programming: The state of the art as seen through the literature review and current studies. Companion paper No. 1 to the Report of the Task Force on Employee Assistance Programs. Addiction Research Foundation, Toronto, 1978.
2. Yankelovich, D. The new psychological contracts at work. *Psychology Today* 11:46–50 (May 1978).
3. Bartell, T. The human relations ideology: An analysis of the social origins of a belief system. *Human Relations* 29:737–749 (1976).
4. National Institute on Drug Abuse (NIDA). *Occupational drug abuse programs.* Final report to NIDA. Rockville, Md., 1977.
5. Stephens, R. C., et al. Drug abuse and the worker: issues in arbitration. University of Houston. Processed.
6. Provost, F., et al. Alcoholism in the workplace: A review of recent arbitration cases. *Employee Relations Law Journal* 4:400–414 (1978).
7. Trice, H. M., and Belasco, J. *Emotional health and employer responsibility.* Bulletin 57. New York State School of Industrial and Labor Relations, May 1966.
8. Roman, P. M. Executive and problem drinking employees. In *Proceedings of the third annual Alcoholism Conference of the National Institute on Alcohol Abuse and Alcoholism*, June 1973.
9. Hilker, R. J., and Asma, F. E. A drug abuse rehabilitation program. *Journal of Occupational Medicine* 17:351–354 (1975).
10. Rogers, R. E., and Colbert, J. T. C. Drug abuse and organizational response: a review and evaluation. *Personnel Journal* 54:266 (1975).
11. Redfield, J. T. Drugs in the workplace—substituting sense for sensationalism. *American Journal of Public Health* 63:1064–1070 (1973).
12. Florida Occupational Program Committee. *Solving job performance problems.* Health and Rehabilitation Services, State of Florida, 1975.
13. Booz, Allen and Hamilton, Inc. A seminar on marketing the occupational

alcoholism program. A report to the National Institute on Alcohol Abuse and Alcoholism. Washington, D.C., September 1975.

14. Cahalan, D., and Cisin, I. H. *Final report on a service-wide survey of attitudes and behavior of naval personnel concerning alcohol and problem drinking.* Washington, D.C.: Bureau of Social Science Research. 1975.
15. Hitz, D. Drunken sailors and others: Drinking problems in specific occupations. *Quarterly Journal of Studies on Alcohol* 34:496–505 (1973).
16. Trice, H. M. *Drug use and abuse in industry.* Washington, D.C.: Office of Drug Abuse Policy, January 1979.
17. Trice, H. M., and Beyer, J. M. Differential use of an alcoholism policy in Federal organizations by skill level employees. In *Employee assistance and alcoholism programs in American industry.* Baltimore: Johns Hopkins Press, 1977.
18. Chambers, C. D. *Differential drug use within the New York State labor force.* Mamaroneck, N.Y.: Starch/Hooperating/The Public Pulse, 1971.
19. Trice, H. M. Drugs, drug abuse and the work place. In *Principles of social pharmacology.* New York: Basic Books, In press.
20. Myrick, R., and Basen, M. *Drug use in industry.* Summary of a final report to NIDA. Rockville, Md.: National Institute on Drug Abuse, 1979.
21. Steele, P. D. Management and union leadership's attitudes concerning drug use in industry. Paper presented at annual meeting of the Society for the Study of Social Problems. New York City, September 1976.
22. Kurtis, C. Drug abuse as a business problem. New York Chamber of Commerce, New York City, September 1979.
23. Halpern, S. *Drug abuse and your company.* New York: American Management Association, 1972.
24. Johnston, R. G. A study of drug abuse among employees in Akron, Ohio. Akron: Bureau of Business and Economic Research, University of Akron, 1971.
25. Rush, H. M. F. Combating employee drug abuse. *Conference Board Record* 8:58–64 (November 1971).
26. Scher, J. M. The impact of the drug abuser on the work organization. In *Drug abuse in industry: growing corporate dilemma.* Springfield, Ill.: Charles C. Thomas, 1973.
27. Urban, M. L. Drugs in industry. In *Drug use in America: Problem in perspective.* Washington, D.C.: National Commission on Marijuana and Drug Abuse, 1972.
28. O'Donnell, J. A., et al. *Young men and drugs: A nationwide survey.* Rockville, Md.: National Institute on Drug Abuse, 1976.
30. Johnston, L. D., et al. *National survey on drug abuse, 1977.* Rockville, Md.: National Institute on Drug Abuse, 1977.
31. Caplovitz, D. The working addict. Graduate School and University of the City University of New York, New York City, 1976.
32. Marijuana: A conversation with NIDA's Robert L. DuPont. *Science* 192:647–650 (1979).
33. Chein, I., et al. *The road to H.* New York: Basic Books, 1964.
34. Devenyi, P., and Wilson, M. Abuse of barbiturates in an alcoholism population. *Canadian Medical Association Journal* 104:219–221. (1971).
35. Threatt, R. M. *The influence of alcohol on work performance.* Raleigh, N.C.: Human Ecology Institute, February 1976.

36. *The Third Special Report to the U.S. Congress on Alcohol and Health.* Washington, D.C.: U.S. Department of Health, Education, and Welfare, 1978.
37. Wrich, J. *The employee assistance program.* Center City, Minn.: Hazelden Foundation, 1974.
38. Gualtieri, P. K., et al. *Typology, classification and evaluation criteria for NIAAA's occupational programs.* Final report of a project to the National Institute on Alcohol Abuse and Alcoholism. Rockville, Md., 1978.
39. National Institute on Drug Abuse. *Developing occupational drug abuse programs: Considerations and approaches.* Rockville, Md., 1978.
40. Roman, P. M. Occupational programming in major American corporations: The 1976 executive caravan survey. NIAAA report for inclusion in the Third Alcoholism and Health Report to Congress of 1977. National Institute on Alcohol Abuse and Alcoholism, Rockville, Md., 1977.
41. DHEW Employee Assistance Program. Report to the Secretary, U.S. Department of Health, Education, and Welfare, Washington, D.C., 1978.
42. Sherman, P. Why so few occupational programs? *Focus* 1:10–11 (February–March 1978).
43. Smith, D., et al. *Report of the Task Force on Employee Assistance Programs.* Toronto: Addiction Research Foundation, 1978.
44. Hilker, R. R. J. A company-sponsored alcoholic rehabilitation program. *Journal of Occupational Medicine* 14:769–772. (1972).
45. Archer, L. D. Statement before the Subcommittee on Federal Spending Practices Committee on Governmental Affairs and Subcommittee on Alcoholism and Drug Abuse, Committee on Human Resources. U.S. Senate, August 17, 1978.
46. Sadler, M., and Horst, J. F. Company/union program for alcoholics. *Harvard Business Review* 50:22 (September–October 1972).
47. Alender, R., and Campbell, T. An evaluation study of an alcohol and drug recovery program: A case study of the Oldsmobile experience. *Human Resource Management* 14:14–18 (Spring 1975).
48. Winslow, W., et al. Some economic estimates of job disruption. *Archives of Environmental Health* 13:213–219 (1966).
49. Substantial cost savings from establishment of alcohol programs for Federal civilian employees. Report to the Special Subcommittee on Alcohol and Narcotics of the Committee on Labor and Public Welfare, U.S. Senate, 1970.
50. Editorial. *British Journal of Addiction* 65:259–261 (1970).
51. Schlenger, W. E., and Hayward, B. J. *Assessing the impact of occupational programs.* Raleigh, N.C.: Human Ecology Institute, 1975.
52. Borthwich, R. B. *Summary of cost-benefit study results of Navy alcohol rehabilitation program.* Technical Report No. 346. Presearch, Inc., contract to the Department of the Navy. Washington, D.C., 1977.
53. Perlin, L. *Drug abuse in industry.* Washington, D.C.: AFL-CIO, 1971.
54. Trice, H. M. Alcoholism programs in unionized work settings: Problems and prospects in union-management cooperation. *Journal of Drug Issues* 7:103–115 (Spring 1977).
55. Hallan, J. B., and Holder, H. D. *Occupational programming: A guide to health insurance coverage for alcoholism.* Report to National Institute on Alcohol Abuse and Alcoholism. Raleigh, N.C., December 1976.

56. Whiton, R. R. Considerations on health insurance benefits for alcohol treatment: An update. Raleigh Hills Hospital, Raleigh, N.C., 1976.
57. *Utilization of third-party payments for financing of drug abuse treatment.* Rockville, Md.: National Institute on Drug Abuse, 1977.
58. Schlenger, W. E., and Hayward, B. J. Occupational programming: Problems in research and evaluation. *Alcohol Health and Research World,* Spring 1976.
59. Bennett, G. K. Nature and extent of employee assistance programs in Ontario and elsewhere. Companion paper No. 7 to the Report of the Task Force on Employee Assistance Programs. Addiction Research Foundation, Toronto, 1978.

28

Smoking Cessation Programs in Occupational Settings

Brian G. Danaher, Ph.D.

Cigarette smoking has been called this country's foremost preventable cause of death and disability and its greatest public health problem *(1, 2)*. The estimated data for premature death and unnecessary disability are staggering. In 1977, smoking played a major role in 220,000 deaths from heart disease, 78,000 lung cancer deaths, and 22,000 deaths from other causes *(1)*. Smoking has been estimated to be responsible for 20 percent of all cancer, 25 percent of all cardiovascular disease, and 40 percent of all respiratory disease *(3)*.

To a considerable extent, public health campaigns have successfully alerted the general public about these risks, and most smokers (perhaps as many as 90 percent) have stated that they would like to quit. A survey by the Center for Disease Control in 1976 and a Gallup poll in 1974 found that upwards of 60 percent of adult smokers had made at least one serious attempt to quit smoking *(4, 5)*. These surveys, however, also showed that almost 53 million Americans still smoked.

The economic costs attributed to smoking have been catalogued and found to be of similar vast proportions. Luce and Schweitzer *(6)* estimate that

This article was first published in *Public Health Reports* (March–April 1980), pages 149–156, and appears here in slightly revised form.

The author is Assistant Professor, Division of Behavioral Sciences and Health Education, School of Public Health, University of California, Los Angeles.

in 1976 cigarette smoking cost the country $27.5 billion, of which $19 billion was attributed to lost production. Estimates of the number of working days lost annually because of smoking range from 77 million in 1971 *(7)* to 81 million in 1978 *(8)*. One source has suggested that smoking costs $3 per day per smoking employee, based on insurance costs, sick days, absenteeism, down time, lost productivity, and maintenance costs *(9)*.

The growing recognition in the business community of the overwhelming evidence of dangerous health consequences from smoking and a clearer perception of its more immediate economic impact have led an increasing number of companies—at least 3 percent of all U.S. companies and 6 percent of all Canadian companies—to offer special programs or incentive plans to encourage employees to stop smoking *(10)*. There has been also a concurrent growth of interest within the scientific community about the potential benefits of systematically encouraging smoking cessation in the occupational setting, interest that has been heightened by epidemiologic evidence showing greatly accentuated risks from occupational cancers related to smoking *(11)*.

In this report I summarize the current status of smoking control from the perspective of (*a*) the research literature on smoking and (*b*) current smoking control programs in occupational settings, citing examples. (The information cited in the examples was obtained by personal interviews with the programs' medical directors.)

TRENDS IN SMOKING CESSATION RESEARCH

An impressive amount of research worldwide has been directed at uncovering effective methods for smoking cessation *(12–14)*. Helpful methods have been identified, but their absolute effectiveness has proved somewhat disappointing. Hunt and Bespalec *(15)* examined 89 studies in the literature and found that more than half of the persons who had stopped smoking by the end of a program subsequently relapsed; the greatest recidivism appeared within the first five weeks of follow-up. These compelling results, along with the corresponding evidence of changes in the smoking rate (see Figure 28.1), have helped to establish a 30 percent abstinence level as the benchmark or frame of reference against which the incremental efficacy of specific smoking cessation programs can be measured.

Social learning theory has been one source of optimism about the future of antismoking programs:

> . . . some progress has been made in terms of treatment effects, research methodology, and the ways in which the probelm is conceptualized. This progress justifies a measure of cautious optimism about the future of the field and, because the social learning approach has resulted in the clearest

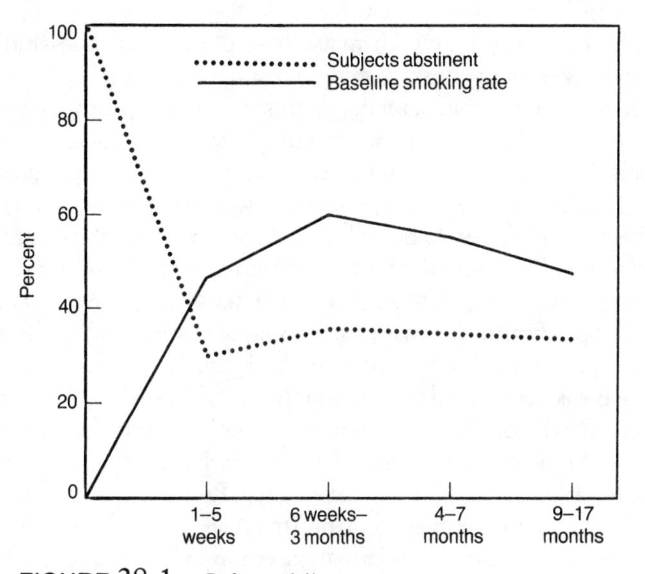

FIGURE 28.1 Relapse following treatment
SOURCE: Chart is adapted from references *13* and *15*.

increments in technical, methodological, and conceptual sophistication, about the fruitfulness of applying it to the modification of smoking behavior. *(12)*

Behavioral Research

Aversive smoking procedures are among the approaches to smoking cessation reported in the literature that seem to hold the most promise. In these procedures, cigarette smoke is used as an unpleasant stimulus—for example, in the form of hot smokey air, oversmoking, and rapid smoking. I consider these approaches here in the context of aversive conditioning, but the importance of pharmacological, cognitive, and self-control features of the experience has also been acknowledged *(13, 16)*.

By far the most research on aversive smoking has focused on the rapid smoking procedure. In the standard format, participants meet alone or in a small group and are instructed to smoke successive cigarettes in an accelerated manner (puff every six seconds) until either a personal tolerance or a time limit is reached (whichever occurs first). In early research, uniform (100 percent) abstinence was reported at termination and 60 percent abstinence at six-month follow-up, fully twice the benchmark level noted earlier. Results from more recent research have proved somewhat less impressive. Overall, however, the rapid smoking procedure has been found to be relatively more effective than other cessation programs. Its absolute effectiveness ranks above the 30 percent level used as a general measure of efficacy.

Unfortunately, rapid smoking cannot be used by many smokers, as it acutely stresses the cardiovascular system. Research is continuing both to determine the extent of risk and to establish a risk-benefit perspective *(16, 17)*. But it seems clear that the careful screening of participants that is required seriously limits the general applicability of the rapid smoking approach.

Less stressful alternatives to rapid smoking are being sought, and a procedure known as regular-paced aversive smoking (RPAS) has received attention. Specifically, in this aversive smoking approach, a slower-paced puffing tempo (one puff every 30 seconds) is taught, but the same context and emphasis is maintained as in the more stressful rapid smoking approach. Initial results indicate that RPAS is as effective as rapid smoking in producing abstinence at program termination. Unfortunately, the relapse observed during long-term follow-up is somewhat greater. Research is continuing to ascertain how various self-management strategies might be combined with RPAS to encourage a more enduring pattern of behavior change.

Aversive approaches to smoking control emphasize only one side of the solution—they help the person to avoid smoking. They do not adequately provide the novice ex-smoker with substitute skills to help him or her actively resist relapse. The development of a repertoire of substitute or nonsmoking skills is a positive approach to becoming an ex-smoker and one that has received considerable research attention under the title of self-control or self-management. One final point, the control of cigarette smoking is perhaps best viewed as an example of self-control, since it is behavior that is approved, modeled, or tolerated by most people, and its deleterious consequences are predominantly distant or removed from immediate experience *(18)*.

A variety of self-control approaches have been devised, emphasizing (*a*) situational controls (for example, managing one's environment to facilitate smoking cessation), (*b*) self-congratulatory statements or cognitions (thoughts) when smoking is controlled, (*c*) punishment for smoking (for example, forfeiture of a financial deposit), and (*d*) reinforcement of alternatives to smoking. Even though these programs have included some of the most innovative methods to date (such as smoking at random intervals cued by pocket timers, the use of locking cigarette cases, and the forfeiture of financial deposits to organizations identified by the smoker as "most hated"), few of these innovations have been found to be effective by themselves *(13)*. A number of these self-control strategies have also been combined in the hope that together they might have a beneficial synergistic effect, but this hoped-for result generally has not been realized *(13, 19)*.

More promising are the so-called broad-spectrum behavioral programs that combine aversive smoking and self-control components. In these programs, the smoker is helped to stop smoking through aversive smoking experiences and then is encouraged to remain an ex-smoker by practicing a set of self-control strategies for overcoming lingering urges to smoke. A number of studies of broad-spectrum programs suggest that this more comprehensive approach may be fruitful *(12, 19)*.

An important, but only lightly investigated, feature of behavioral programs for smoking control is the format used for delivering program instructions. In most self-control programs, a manual provides step-by-step instructions while also serving as a permanent guide. Although the efficacy of programs using a manual and self-help generally has been incompletely documented, this line of program development is viewed as critically important *(20)*. Moreover, the availability of comprehensive behavioral programs in manual form—for example, those of Danaher and Lichtenstein *(19)* and of Pomerleau and Pomerleau *(21)*—facilitates the implementation of programs based on our current research knowledge.

Other communications media have also been developed for delivering smoking control program instructions. McAlister *(22)* found that a videotaped smoking control program that served as the basis for a class directed by a paraprofessional was as effective (at least in the short run) as a more costly quit clinic directed by a trained consultant. Danaher and associates *(23, 24)* found that audiotapes for practicing relaxation and for regular-paced aversive smoking could be used to reduce the contact time of participants with program staff. Finally, Dubren *(25)* reported that a telephone answering system could be programmed to provide maintenance messages to participants in a previous quit clinic.

As noted earlier, these programs using various communications media can often be applied as adjuncts in a traditional quit clinic or class context and will reduce unnecessary contact time with the program staff. More exciting is the fact that such approaches can be used in innovative self-help formats in which professional contact time is held to an absolute minimum. In both cases, programs using communications media materials can reduce overhead and improve the cost-effectiveness of smoking cessation programs, as has been urged by Green and associates *(26)*. Perhaps most significant is the fact that media materials can be used to reach out to those smokers in the population who would not attend a smoking clinic *(27)*.

Physician Counseling

Interventions by personal physicians have been suggested for a variety of health behavior problems. The rationale is usually that health risk information and a few carefully chosen suggestions from a physician can be very effective. Lichtenstein and Danaher *(28)* have provided a useful schema for examining the physician's potential role in encouraging smoking cessation:

for example, acting as a model, supplying information, issuing admonishments, facilitating referrals, or directly managing an intensive program for smoking cessation.

Pincherle and Wright *(29)*, who examined the efficacy of physician advice delivered during annual physical examinations sponsored by businesses, found that only 13 percent of the employees given these examinations quit smoking; 7 percent actually started smoking or increased their smoking rate in the period following the examination. Meyer and Henderson *(30)* found that a brief discussion with a physician about the risk of cardiovascular disease from smoking resulted in only 3 of 14 employees abstaining from cigarettes at follow-up. These authors found, however, that discussion with a physician was as effective as more intensive counseling focused on modification of multiple cardiovascular risk factors.

The best results with physician advice in smoking control have been associated with smokers who have recently experienced a myocardial infarction. The least impressive results are traced to programs that have been directed at persons taking part in routine physical examinations without having any presenting complaint *(28)*. Compared with more intensive approaches, physician counseling will probably produce less impressive absolute changes in smoking behavior. However, its cost-effectiveness should recommend it in many settings, and it should provide useful assistance to a great number of smokers *(31)*.

Proprietary Program

Commercial interest has grown in setting up fee-for-service smoking control programs targeted at the general community and occupational settings. Representative examples of these proprietary or commercial programs have been reviewed comprehensively by Schwartz and Rider *(14)*. If we take this source as a guide, we find that by far the greatest effort to date in this area has come from SmokEnders *(32)*. The SmokEnders program follows a highly structured format in which a gradual reduction in smoking is encouraged, followed by sessions devoted to reinforcement. References in the SmokEnders' promotional literature and statements by J. Rogers, the organization's co-founder [cited by Schwartz and Rider *(14)*], claim a success rate of almost 90 percent. In more objective assessments reported in the literature *(33)*, less than one-half the claimed level of long-term success was found. Schwartz and Rider *(14)*, who examined these data more conservatively, reported a 27 percent abstinence level. For the most part, proprietary programs have followed a group treatment model and have used the strategies outlined in the research literature. Unfortunately, these programs have also displayed a reluctance to permit any careful, outside evaluation of their effectiveness.

Nonprofit Programs

A number of community organizations and foundations have participated in smoking control programs. Perhaps the most widely known is the quit clinic provided through the American Cancer Society. The ACS program, which has been offered to thousands of smokers, follows a three-step sequence or approach: (*a*) self-appraisal and insight development, (*b*) practice in abstinence under controlled conditions, and (*c*) a maintenance phase. In one of the few existing outside evaluations (*34*), 29 ACS clinics were studied. Abstinence rates based on all participants were 30 percent at 6 months after the programs, 22 percent at 12 months, and 18 percent at 18 months.

CURRENT OCCUPATIONAL SMOKING CONTROL PROGRAMS

Physician Counseling

A number of businesses have chosen to emphasize the physician counseling approach to smoking cessation. That is, at annual or biennial company-sponsored physical examinations the physician is encouraged to comment on the employee's need to quit smoking and then to offer personal advice or suggestions to the employee. E. I. DuPont De Nemours, for example, has planned to use this approach with its employees, perhaps supplementing it with booklets and filmstrips. One significant and often overlooked advantageous feature of the physician model is that it is relatively unobtrusive. Because the health and medical recommendations of the physician may not be perceived by the employee as being an order directly from management, the employee may be less inclined to regard them as as intrusion into his or her personal life and habits.

Outside Consultants

A large number of businesses are sponsoring outside consultant groups to help their employees stop smoking. Some have simply announced the availability of smoking cessation programs, while others have even helped to defray some or all of the costs of participation in such programs. For example, Eastman Kodak and parts of Western Electric, General Foods, and Xerox have used this referral approach. Others have invited consultants into the business setting to conduct on-site programs. Companies following the consultant-at-the-work-site model include Campbell Soup, American Telephone and Telegraph, Johns-Manville, and Boeing Aircraft. In many cases, the consultant group has been SmokEnders, although Seventh-Day Adventist and American Cancer Society groups have also been used.

Perhaps because the use of outside consultants removes a sense of personal responsibility for program evaluation, few businesses have systematically examined the efficacy of these approaches beyond informal personal reactions and anecdotal feedback. However, there are exceptions.

Example. The Campbell Soup Company (Camden, N.J.) developed a working agreement with the Center for Behavioral Medicine at the University of Pennsylvania to conduct a series of smoking cessation classes. The consultant in this instance was a seasoned clinical assistant who had prior experience in programs based on recently published core materials *(21)*. The model was one of providing a service, collecting data, and training the staff at the business so that in-house programs could be carried out. Participants and the company split the fee of $50 ($20 to $30). To date, three small groups (N = 36) have been organized and at the six-month followup, 25 percent of participants were found to be abstinent. Additional programs are planned.

Example. Boeing Aircraft sponsored an in-house program, conducted by the Seventh-Day Adventist Church, for 35 people. Questionnaire results obtained from 27 participants (77 percent) showed that 50 percent were abstinent at the end of treatment, while only 30 percent were not smoking at the follow-up at three or more months.

In-House Programs

Smoking cessation programs can be offered as part of a company's health education or occupational health program. It is in discussion of the approaches taken in these in-house programs that the greatest variety and innovation can be found. The available reports can be roughly grouped as describing (*a*) group training in skills and educational programs, (*b*) incentive programs, and (*c*) prohibitions on smoking.

Group skills and educational programs usually follow a format similar to that provided by the quit clinics (both commercial and voluntary organizations). A number of businesses have offered employees quit clinics (for example, the Ford Motor Company and the Campbell Soup Company), and a number are contemplating an expansion of their smoking cessation activities to include in-house programs (Boeing Aircraft Company and the Xerox Corporation).

Example. The Ford Motor Company World Headquarters began a systematic effort to set up an effective in-house smoking cessation program through its corporate health education program. A pilot study was undertaken to test the efficacy of various self-help formats. Approximately 40 percent of the smokers involved in the corporate cardiovascular risk intervention program were invited to participate. Self-help groups were formed composed of colleagues and friends who shared an interest in using one of three methods: regular-paced aversive smoking and self-control (26 persons in six groups), abrupt withdrawal with contingency contracting and self-control (22 persons in four groups), or gradual withdrawal using contingency contracting and self-control (10 persons in four groups). Each group elected a leader who helped see that the weekly agenda topics were covered and that subsequent meetings were scheduled. The materials for the program provided by Ford were in workbook form and included strategies and procedures from the behavioral program outlined by Danaher and Lichtenstein *(19)*. Audiotapes also were used to increase the cost-effectiveness of the program.

Partly because of the participants' limited contact with Ford's health education office, unexpected difficulties were encountered in collecting data. At termination of the program, the greatest reduction in smoking was found among persons who had been in groups using the aversive smoking approach. At a six-month follow-up assessment, 20 percent of this program's participants were not smoking.

A number of conclusions were drawn from this study: namely, that (*a*) a wider range of smoking cessation opportunities should be made available, including face-to-face quit clinics; (*b*) more preparation and effort are required to accomplish a more satisfactory data analysis; and (*c*) more efficient operation would be achieved by offering a structured readiness program that would allow employees to examine their personal smoking habits without the pressures to abstain. These expanded plans were to be tested in 1979 at another Ford Division plant.

Incentive programs, that is, smoking cessation programs in which company-sponsored monetary awards are systematically used to encourage nonsmoking, have proved popular. Recent reports appearing in national publications such as the *Wall Street Journal (35)*, the *Los Angeles Times (36)*, and *Business Week (6)* have indicated widespread use of, and apparent success with, incentive programs. To date, the majority of programs of this type appear to have been sponsored by smaller businesses, for example, an ambulance company *(37)* and a cosmetics firm *(36)*. Criteria for rewards can be either not smoking while at work or more complete abstinence. A careful search failed to uncover any programs in which an incentive approach was combined with opportunities for training in nonsmoking skills.

Example. The Texas operating division of the Dow Chemical Company instituted an innovative smoking cessation program after it discovered that workdays lost by smokers cost the company an estimated $500,000. A one-year program was undertaken in which lotteries and financial awards were used. In one lottery aimed at smokers, every month of nonsmoking earned the novice ex-smoker one chance to win a boat and motor valued at $2,400. In the other, strictly monetary contest, weekly $1 bonuses were offered for abstinence, as well as a chance to win a $50 bonus quarterly. The second lottery was aimed at recruiters, employees who would encourage smokers to join the quit program. A recruiter earned one chance toward a boat and motor prize for every month of nonsmoking reported by a recruit.

Almost 400 employees (24 percent of the smokers) were recruited, and at the end of the program an impressive 76 percent were abstinent. Even though enthusiasm over these results must be tempered by the lack of follow-up data and objective measures that would corroborate self-report, the Dow Chemical Company program illustrates how incentives can be used to promote both recruitment and significant changes in smoking behavior.

A blanket prohibition disallowing all smoking at work is the most restrictive approach to smoking cessation in the occupational setting. In some cases these prohibitions stem from historical precedent as was the case with the Campbell Soup Company *(10)*. Other, more recent instances of the prohibition approach are based on evidence of an association between certain occupations and smoking behavior *(38)* and on the synergistic effect of smoking and occupational exposure on cancer *(11)*.

It should be noted that complete prohibition of smoking need not be the only approach used when there is potential exposure of workers to occupational carcinogens. The Tyler Asbestos Workers Program, for example, emphasized both health education and physician counseling *(39)*.

Enlightened businesses that prohibit smoking also appear to provide employees easier access to special smoking cessation programs. For example, the Johns-Manville Company prohibits all smoking because of potential asbestos exposure, but it follows a policy of first offering opportunities for its employees to attend SmokEnders clinics. Effective smoking cessation programs are complementary to—not substitutes for—efforts to reduce environmental exposure to harmful ingredients in the workplace.

Example. Johns-Manville initiated its aggressive antismoking drive by banning smoking in two plants, one in Massachusetts and one in Texas. Violation of the ban occurred at both sites and produced disciplinary actions. In both instances, the local union filed a grievance, and arbitration decided in favor of the employee in one case and in favor of the company in the other. From experience gained in these early encounters, the Johns-Manville

Company has developed a five-point approach to converting plants from smoking to nonsmoking status: (*a*) health information, (*b*) meetings between local management and union representatives, (*c*) presentation directly to employees of the rationale for not smoking, (*d*) encouragement for workers to attend partially subsidized smoking cessation classes, and (*e*) institution of the smoking ban. In another company decision, a hiring policy requires all future workers to be nonsmokers. The ban was scheduled to be in effect by 1979 in all plants and to encompass a work force of about 8,000 employees *(40)*.

Restrictions or reinforcement plans aimed at encouraging employees to stop smoking at work may not produce all of the results desired. People can learn to manage their smoking so that it occurs only outside of work settings. In fact, Meade and Wald *(41)* recently cited data for more than 2,000 workers in a British food processing factory showing that when prohibited from smoking at work, they made up for lost time by smoking more during other hours. These authors concluded that "it is obviously possible that restrictions on smoking at work may influence total daily consumption, but our data provide no clear evidence of this." Risks from second-hand smoke may be relieved by smoking restrictions *(42)*, but there may not be as significant a beneficial effect on the risk of premature morbidity and mortality or any reduction in workdays missed.

RECOMMENDATIONS FOR CESSATION PROGRAMS

From the research literature cited earlier in this report, we know that even when a participant achieves success in a smoking cessation program, the odds are 7 out of 10 that he or she will relapse before three months have passed. The meager results available for the programs that have been offered in the business setting do not appear to diverge dramatically from this benchmark level. In making recommendations, then, consideration needs to be given to the empirical perspective fostered by social learning theory, which has proved useful in research on smoking behavior.

The empirical perspective, of course, demands the thorough collection and analysis of evidence. In most cases, data are difficult to obtain when programs are turned over to outside consulting firms; this problem may be most acute when the business deals with proprietary programs. The needed research and development in this burgeoning field will require access to outcome data—possibly best accomplished through in-house approaches.

Too often, skills-training programs in occupational settings appear to be based on models that are imported from the smoking clinic, and they are only minimally sensitive to the unique features of the business setting. As Chesney and Feuerstein correctly note *(43)*, on-site programs in business settings reduce the considerable personal costs (lost time and transportation difficulties) associated with clinic-based programs while allowing more opportunity for careful monitoring and follow-up than off-site programs.

Contingency or incentive programs have apparently taken advantage of the unique features of the work setting. However, the omission or underemphasis of skills-training opportunities shows that these programs are based on the naive assumption that smokers require only added motivation to be able to quit. What appears to be needed is careful examination of motivational and skills training approaches, both separately and in combination.

As mentioned earlier, occupational programs present a unique opportunity to maintain contact with people over an extended period, and continuity of contact may be an important ingredient in facilitating enduring abstinence *(43)*. Informal group meetings might be made available to participants who have completed prior programs but find, for whatever reason, that they need to renew their enthusiasm for not smoking. Moreover, these maintenance meetings could be open to all novice ex-smokers in the work force, no matter where or how they achieved cessation. (Availability of a lending library of printed and audiotaped materials would be one cost-effective step in this direction.)

The survey data strongly suggest that most smokers prefer to quit without any formal help. A comprehensive program, then, should be designed not only to help people quit but also to help all workers resist relapse. Antirelapse programs would include behavioral self-control methods for overcoming lingering urges to smoke *(19)*.

Another innovative direction deserving of more attention is the comprehensive wellness approach. In this expanded context, smoking becomes just one element in a multicomponent program aimed at stress reduction, blood pressure management, exercise enhancement, diet control, weight management and so forth. The multitarget approach assists in recruitment of participants for smoking control programs and may be uniquely suited to encouraging the adoption of a repertoire of nonsmoking substitute behaviors *(44)*.

Recruitment of participants for smoking cessation programs remains a largely unexplored area of key importance. The optimal manner of announcing or publicizing smoking cessation programs has not been determined. One promising idea used by the Campbell Soup Company is to hold a health fair in which attention is focused on many aspects of personal health. Recruitment for smoking cessation programs might be enhanced if there were a way to provide some personal assessment of a person's risk status, as in cardiovascular risk *(30)*, through the Health Hazard Appraisal method *(45)*, which the

Centers for Disease Control of the Department of Health and Human Services planned to use in an employee program, or through providing employees feedback information on carbon monoxide tests, as was done at a Blue Cross facility. Recruitment might also be greater if a variety of smoking cessation methods were made available, including group clinic programs, self-help groups, and individual self-help materials. The availability of attractive and effective smoking cessation programs over a period of several years would probably improve recruitment simply because of word-of-mouth communication *(46)*. Readiness programs similar to the one noted in the discussion of the Ford Motor Company's program might be another way to involve people who wish to assess their own personal interest in participating in a smoking cessation program. Finally, the incentive approach used by the Dow Chemical Company could encourage a more exhaustive recruitment effort.

A final, and I hope, a resounding point is that additional attention must be focused on careful program evaluation. Even though thousands of programs are apparently being devised for the control of smoking in occupational settings, I found only a few examples of careful evaluation of such programs. Program evaluations must include data on the number of persons recruited (with the percentage of the employee group that is of interest), the proportion of these recruits who have completed the program and their performance (that is, percentage who are abstinent), and the success with which this group maintains nonsmoking status over the course of follow-up (at least six months after completion of the program). Follow-up assessments require preparation and resources if they are to be complete records of outcome. They are a necessary ingredient of a data-based program. Self-reports, particularly under the conditions found in incentive programs, require validation through the available chemical tests of expired air carbon monoxide or plasma thiocyanate *(47)*.

Manpower for programs in occupational settings can be obtained from a number of sources. Boeing Aircraft Company, for example, has established a program in which University of Washington seniors majoring in health studies participate in various aspects of the planning and implementation of smoking control programs. Graduate students from schools of public health represent another talented, but largely untapped, resource.

The programs used as examples in this report might be thought of as prototypes. As with all prototypes, they are rough approximations of the final product. The process of refinement and development requires testing (data collection) and replication. Certainly there will have to be greatly expanded support for the assessment, as well as for the innovative development, of prototype smoking cessation programs. A report on the status of smoking cessation programs five years hence will have to include considerably more evidence as to the effectiveness of competing models and about the criteria needed for determining the most appropriate match between a program and a business if the occupational setting is to fulfill its promise as an exciting arena for carrying out meaningful preventive health programs.

REFERENCES

1. Califano, J. A., Jr. Keynote address before the National Interagency Council on Smoking and Health, Washington, D.C., January 11, 1978. Department of Health, Education, and Welfare, Washington, D.C.
2. Center for Disease Control. *The health consequences of smoking: A reference edition.* DHEW Publication No. (CDC) 78–8357. Atlanta, 1976.
3. Boden, L. T. The economic impact of environmental disease on health care delivery. *Journal of Occupational Medicine* 18:467–472 (1976).
4. Center for Disease Control. *Adult use of tobacco 1975.* Atlanta, 1976.
5. Public puffs on after ten years of warnings. *Gallup Opinion Index* no. 108:20–21 (June 1974).
6. Luce, R. B., and Schweitzer, S. D. Smoking and alcohol abuse: a comparison of their economic consequences. New England Journal of Medicine 298:569–571 (1978).
7. Terry, L. T. The future of an illusion. *American Journal of Public Health* 61:233–240 (1971).
8. American Cancer Society. A *national dilemma: cigarette smoking* or *health of Americans.* Report of the National Commission on Smoking and Public Policy. American Cancer Society, New York, 1978.
9. Mattes, B., Services Division, SmokEnders Corp., Phillipsburg, N.J. Personal communication, October 1978.
10. Companies put up the "no-smoking" sign. *Business Week* no. 2536:68 (May 29, 1978).
11. Hoffman, D., and Wynder, E. L. Smoking and occupational cancers. *Preventive Medicine* 5:245–261 (1976).
12. Bernstein, D. A., and McAlister, A. L. The modification of smoking behavior: programs and problems. *Addictive Behaviors* 1:89–102 (1976).
13. Lichtenstein, E., and Danaher, B. G. Modification of smoking behavior: A critical analysis of theory, research, and practice. In *Progress in behavior modification.* vol. 3, edited by M. Hersen, R. M. Eisler, and P. M. Miller. New York: Academic Press, 1976.
14. Schwartz, J. L., and Rider, G. *Review and evaluation of smoking control methods: The United States and Canada, 1969–1977.* DHEW Publication no. (CDC) 79–8369. Atlanta: Center for Disease Control, 1978.
15. Hunt, W. A., and Bespalect, D. A. An evaluation of current methods of modifying smoking behavior. *Journal of Clinical Psychology* 30:431–438 (1974).
16. Danaher, B. G. Research on rapid smoking: interim summary and recommendations. *Addictive Behaviors* 2:151–166 (1977).

17. Sachs, D. P. L., et al. Clarification of risk-benefits issues in rapid smoking. *Journal of Consulting and Clinical Psychology* 47:1053–1060 (1979).
18. Thoresen, C. E., and Mahoney, M. J. Behavioral self-control. New York: Holt, Rinehart and Winston, 1974.
19. Danaher, B. G., and Lichtenstein, E. *Become an ex-smoker.* Englewood Cliffs, N.J.: Prentice-Hall, 1978.
20. Glasgow, R. E., and Rosen, G. M. Behavioral bibliography: A review of self-help behavior therapy manuals. *Psychological Bulletin* 85:23 (1978).
21. Pomerleau, O. F., and Pomerleau, C. S. *Break the smoking habit: A behavioral program for giving up cigarettes.* Champaign, Ill., Research Press, 1977.
22. McAlister, A. L. Toward the mass communication of behavioral counseling: A preliminary experimental study of a televised program to assist in smoking cessation. Doctoral dissertation. Stanford University, 1976.
23. Danaher, B. G., et al. A smoking cessation program for pregnant women: an exploratory study. *American Journal of Public Health* 68:896–898 (1978).
24. Danaher, B. G., et al. Aversive smoking using printed instructions and audiotape adjuncts. *Addictive Behaviors* 5:353–358 (1980).
25. Dubren, R. Self-reinforcement by recorded telephone messages to maintain non-smoking behavior. *Journal of Consulting and Clinical Psychology* 45:358–360 (1977).
26. Green, L. W.; Rimer, B.; and Bertera, R. How cost-effective are smoking cessation methods? In *Progress in smoking cessation; Proceedings of the International Conference on Smoking Cessation, New York, June 1978,* edited by J. L. Schwartz. New York: American Cancer Society, 1978.
27. Schwartz, J. L., and Dubitsky, M. Expressed willingness of smokers to try 10 smoking withdrawal methods. *Public Health Reports* 82:855–861 (1967).
28. Lichtenstein, E., and Danaher, B. G. What can the physician do to assist the patient to stop smoking? In *Chronic obstructive lung disease: Clinical treatment and management,* edited by R. E. Brashear and M. L. Rhodes. St. Louis: C. V. Mosby Co., 1978.
29. Pincherle, F., and Wright, H. B. Smoking habits of business executives: doctor variation in reducing cigarette consumption. *Practitioner* 205:209–212 (1970).
30. Meyer, A. J., and Henderson, J. B. Multiple risk factor reduction in the prevention of cardiovascular disease. *Preventive Medicine* 3:225–236 (1974).
31. Russell, M. A. H., et al. Effect of general practitioners' advice against smoking. 2:231–235 (1979).
32. Rogers, J. *You can stop: A SmokEnder approach to quitting smoking and sticking to it.* New York: Simon and Schuster, 1977.
33. Kanzler, M.; Jaffe, J. S.; and Zeidenberg, P. Long- and short-term effectiveness of a large-scale proprietary smoking cessation program: a four-year follow-up of SmokEnders participants. *Journal of Clinical Psychology* 32:661–674 (1976).
34. Pyszka, R. H.; Ruggels, W. L.; and Janowicz, L. M. *Health behavior change: Smoking cessation.* Independent Research and Development Report. Menlo Park, Calif.: SRI International, December 1977.
35. Kelliher, E. V. Fewer workers now are singing "smoke gets in your eyes." *Wall Street Journal,* November 7, 1978, pp. 1, 33.

36. McDougall, A. K. Smoking ban at workplace a fiery issue. *Los Angeles Times*, September 1, 1978, pp. 1, 21, 22, 23.
37. Rosen, G. M., and Lichtenstein, E. A workers' incentive program for the reduction of cigarette smoking. *Journal of Consulting and Clinical Psychology* 45:957 (1977).
38. Sterling, R. D., and Weinkam, J. J. Smoking characteristics by type of employment. *Journal of Occupational Medicine* 18:743–754 (1976).
39. Ellis, B. H. How to reach and convince asbestos workers to give up smoking. In *Progress in smoking cessation; Proceedings of the International Conference on Smoking Cessation, New York, 1978,* edited by J. L. Schwartz. New York: American Cancer Society, 1978.
40. Johns-Manville, Inc. (Health, Safety and Environment Department). *Non-smoking program: Policies, program history, implementation, educational material.* Denver, 1978.
41. Meade, T. W., and Wald, N. J. Cigarette smoking patterns during the working day. *British Journal of Preventive and Social Medicine* 31:28 (1977).
42. Shimp, D. M.; Blumrosen, A. W.; and Finifter, S. B. *How to protect your health at work.* Salem, N.J.: Environmental Improvement Associates, 1976.
43. Chesney, M. A., and Feuerstein, M. Behavioral medicine in the occupational setting. In *Behavioral medicine,* edited by J. McNamara. Kalamazoo, Mich.: Behaviordelia, 1979.
44. Cathcart, L. M. A four year study of executive health risk. *Journal of Occupational Medicine* 19:354–357 (1977).
45. Robbins, L. C., and Hall, J. H. *How to practice prospective medicine.* Indianapolis: Methodist Hospital of Indiana, 1970.
46. Nickerson, H. H. An evaluation of health education programs in occupational settings. *Health Education Monographs* (San Francisco) no. 22, vol. 1, 1967, pp. 16–31.
47. Vogt, T. M., et al. Expired air carbon monoxide and serum thiocyanate as objective measures of cigarette exposure. *American Journal of Public Health* 67:545–549 (1977).

29

Stress Management in Occupational Settings

Gary E. Schwartz, Ph.D.

Substantial progress in documenting the role of psychosocial stress in the etiology and development of physical and mental disease has been made in the past 10 years. It is now known that not only are the classic psychosomatic disorders—such as hypertension, ulcers, and asthma—strongly influenced by psychosocial stress, but even susceptibility to and recovery from infectious and genetic disorders (ranging from the common cold to cancer) are determined, at least in part, by stress in the environment and the person's way of coping with stress *(1, 2)*.

Also in the past decade, substantial progress has been made in documenting effective behavioral approaches to the management of psychological and physiological responses to stress. We now know that various behavioral techniques including relaxation, meditation, biofeedback, and other psychological self-control procedures can be helpful in treating some persons with mental and physical disorders *(3)*. These behavioral techniques, when integrated, within a comprehensive, biobehavioral approach to health and illness, can also (*a*) enhance the effectiveness of biomedical treatments such as drugs and, in the process, reduce the dosage needed to produce a given

This article was first published in *Public Health Reports* (March–April 1980), pages 99–108, and appears here in slightly revised form.

Dr. Schwartz is a professor in the Department of Psychology, Yale University, and in the Department of Psychiatry, Yale University School of Medicine.

clinical effect; and (*b*) decrease susceptibility to future disease by promoting positive approaches to health. In other words, stress management procedures are not only useful in treatment, but they also may be useful in the prevention of illness.

It should be recognized at the outset that most of the research conducted to date that links stress to illness is not based on data derived from occupational settings. Although a growing body of literature specifically documents the effects of occupational stress in the etiology and development of physical and mental disease *(4, 5)*, data that directly test the effects of stress management procedures in the treatment and prevention of stress-related disorders in occupational settings are scarce *(6)*. The relative lack of data in this area is neither surprising nor discouraging, because it is only recently that basic and clinical research on stress has developed to the point of seriously justifying research and applications in occupational settings. Furthermore, industry just recently became aware of (*a*) its role in promoting and sustaining health and (*b*) the potential benefits that may accrue by industry, labor, and science working together to promote health in occupational settings.

The challenge facing industry, labor, and science today is to design stress management programs that can be both clinically effective and cost-effective and then to carefully evaluate these programs in occupational settings through systematic research. The promise of positive results emerging from collaborative research in this area is substantial and should be pursued vigorously, despite the numerous problems in conducting such research.

We are witnessing today a major change in our conception of health and illness. In the past, psychological and biological models of health and illness were couched in separate scientific languages and practiced by separate disciplines—now these separations are being broken down. Behavioral and biomedical sciences are beginning to join forces to tackle health problems that require a multidisciplinary approach to their solution. The concept of stress and its implications for health and illness is a key factor bringing these disparate disciplines together.

The extent of this change in orientation can be seen, for example, in the emergence of the new field of behavioral medicine. Formally established at the Yale Conference on Behavioral Medicine in February 1977 *(7)* and extended at a meeting hosted by the Institute of Medicine of the National Academy of Sciences in April 1978, which founded the Academy of Behavioral Medicine Research, Behavioral Medicine has been defined as follows:

> Behavioral Medicine is the interdisciplinary field concerned with the development and integration of behavioral and biomedical science, knowledge, and techniques relevant to health and illness and the application of this knowledge and these techniques to prevention, diagnosis, treatment and rehabilitation. *(8, 8a)*

Two words in this definition, "development" and "integration," need to be emphasized, because they highlight the pitfalls and promise of stress management as applied to health and illness in occupational settings. Although the present data are encouraging, many key basic and applied questions still need to be answered. The development of this knowledge will hinge on the creative integration of behavioral and biomedical approaches. When applied to the occupational setting, the development of this knowledge will further hinge on the integration and collaboration of industry, labor, and science. As Neal Miller, Ph.D., one of the pioneers of behavioral approaches to health and illness, said concerning behavioral medicine's future, we must be "bold in what we try, but cautious in what we claim." In this spirit, I will review what is and what is not known about stress management in occupational settings, suggest some of the directions to be considered for future research and applications, and indicate the cautions that must be considered in light of our present state of knowledge.

The "Models Linking Stress to Illness" are presented on pages 246–249 for those who are not familiar with the psychobiology of stress and illness.

The literature linking occupational stress to disease is reviewed in a 1974 volume *(4)* and in the proceedings of a 1977 conference sponsored by National Institute for Occupational Safety and Health (NIOSH) *(5)*. Some of the major classes of psychosocial stress facing workers have been summarized recently by James S. J. Manuso, Ph.D., director of the Emotional Health program, Equitable Life Assurance Society:

1. Work overload, or work stagnation
2. Extreme ambiguity, or rigidity in relation to one's tasks
3. Extreme role conflict, or little conflict
4. Extreme amounts of responsibility (especially for other people), or little responsibility
5. Cut-throat and negative competition (or one-upmanship), or no competition
6. Constant change and daily variability, or a deadening routinized stability
7. Ongoing contact with "stress carriers" (e.g., demanding workaholics, highly anxious people, indecisive individuals), or social isolation
8. That the corporation, for its own survival, encourages its employees to define their egos in terms of the organization, to contain emotional reactions, and to depend upon it, and
9. The interaction of one's stage of career development, career opportunity, and management style. *(9)*

According to Manuso, "It is not surprising, then, that 58% of the men and 36% of the women in a sample of 95 Emotional Health Program participants at the Equitable Life Assurance Corporation stated that job-related factors, at least in part caused or contributed to their problems" *(9)*.

A recent paper by Chesney and Feuerstein *(10)* highlights some important research on sources of stress. For example, using a homogeneous population of 1,540 white-collar workers (84 percent male) in a large financial institution, Weiman *(11)* examined Selye's *(12)* hypothesis that both overstimulation and understimulation are sources of stress and are associated with a higher level of disease or risk. Weiman confirmed this hypothesis, observing a U-shaped relationship between stimulation (measurement by an index of workload, role conflict, task ambiguity, and responsibility) and an index of disease or risk (including smoking, hypertriglyceridemia, hypercholesterolemia, exogenous obesity, and peptic ulcer). It is of considerable interest that both over- and understimulation can result in an increase in stress-related disorders and behaviors associated with health risk. Chesney and Feuerstein *(10)* comment that research on the health of occupational groups whose jobs are characterized by understimulation, such as blue-collar assembly workers, would further establish this important U-shaped relationship between environment and disease.

Zorn and co-workers *(13)*, in a study of West German sea pilots, observed excess cardiac mortality in this occupational group compared to the cardiac death rate of the male population of Hamburg. Although numerous studies report a relationship between stress and heart disease, the mechanisms linking these two factors remain unknown. To explore the hypothesis that increased catecholamine levels contribute to the relationship between job stress and cardiac death, Zorn and co-workers measured urinary catecholamines in five sea pilots before, immediately after, and 24 hours after a stressful river piloting operation. They found a significant elevation in catecholamines between the pre- and posttrip collections and a subsequent drop in catecholamines 24 hours after the operation.

A related study linking catecholamines and job stress was recently conducted by Dutton and co-workers *(14)*, who compared a group of paramedics with a group of firefighters. Although both groups had similar scores on the Schedule of Recent Life Events—a general life stress scale that predicts susceptibility to disease *(15)*—the paramedics scored significantly higher than the firefighters on a job stress questionnaire designed specifically for the study. The paramedics, importantly, also had significantly higher levels of epinephrine and norepinephrine on work days than on nonwork days.

Chesney and Feuerstein *(10)* recognized that although these studies suggest an association between environmental stress and disease, certain cognitive, personality, and behavior characteristics of the employee (mediated by the brain) interact with characteristics of the environment and influence this association. In collaboration with Chadwick *(16)*, they attempted to define the relationships between job and life stresses, personality characteristics and behavior patterns, job and home environments, physiological strain variables, and coronary heart disease risk and status; they assessed these variables over a 1½-year period for 397 men who were examined at their worksites. The data indicate, for example, that higher levels of catecholamines correlated with job stress as measured by the work pressure subscale of the Work Environment Scale *(17)* and impulsiveness as measured by the Eysenck Personality Inventory *(18)*. Although not mentioned in their report, the implication is that persons high in job stress and high in impulsiveness will more likely evidence health risk factors than persons high in either one alone. As discussed in the "Models Linking Stress to Illness," the need for multimeasure, interactive analyses is critical if the effects of job stress on health are to be understood and therefore controlled.

It must be recognized that combinations of factors within and outside the work situation interact and contribute to disease. Because the work setting either may be a primary determinant of risk or may interact with serious stresses in the worker's personal life, the study of the relationship between job stress and illness is complicated. On the other hand, the control role that the work situation plays in people's lives increases the potential impact that industry can have in motivating persons to change their life-styles for the sake of their health. Industry may, for its own purposes, wish to reduce absenteeism, enhance productivity, and reduce insurance and medical costs. However, providing stress management training as part of a more comprehensive health enhancement program not only may help the worker in the work situation, but also may help the worker to deal with significant problems occurring outside the work situation. In this way, industry can potentially make a greater contribution to society at large.

INTRODUCTION TO STRESS MANAGEMENT PROCEDURES

Numerous procedures are documented by research that can influence response to stress. Some procedures are geared toward helping people change their environment to be more healthful. For example, various studies document how assertiveness training can be used to help people take better control of their lives and in the process reduce tension and hence decrease the physiological responses of strain due to excessive anger or anxiety, or both *(19)*. The goal of assertiveness training programs is not to make people more aggressive, but rather to help them to assert themselves more appropriately in order to reduce the likelihood that they will engage in health risk behaviors reinforced by peer pressure, or to help them modify their jobs (through appropriate channels) to be more healthful. Often, assertiveness training programs consist of combinations of behavior therapy, imagery, role playing, and other techniques aimed at improving people's ability to communicate their concerns, which helps them change the groups in which they work—

not only to better meet their individual needs, but also to improve the functioning of the group as a whole.

Other stress management procedures are geared toward helping people cope with an environment that cannot be changed. These coping procedures involve various mental and psychophysiological techniques, including relaxation, meditation, biofeedback, and guided imagery. For example, progressive relaxation involves teaching people to tense and relax each of the major muscle groups of the body—a "somatic" procedure *(20)*, while autogenic training involves teaching people to imagine particular sensations—such as one's limbs being warm and heavy—a "cognitive" procedure—with the goal of reducing autonomic arousal *(21)*.

Other techniques combine various mental and somatic relaxation procedures. The most well known of these procedures was developed by Benson *(22)*, who proposes that the harmful effects of prolonged psychosocial stress are mediated by excessive elicitation of a hypothalamically controlled "fight or flight" response, with its attendant increased sympathetic nervous activity. Benson further proposes that a reaction opposite in its physiological effects to those of the "fight or flight" response is an integrated "relaxation response" also mediated by the hypothalamus. The relaxation response is presumably elicited by a variety of relaxation and meditation techniques. Goleman and Schwartz *(23)* also have documented the effectiveness of relaxation response procedures.

Benson's technique draws on a combination of processes to promote the relaxation response. It includes (*a*) relaxing all skeletal muscles, (*b*) paying attention to breathing in a relaxed fashion, (*c*) saying a simple word ("one") after each breath (to help remove distracting thoughts), and (*d*) adopting a passive attitude (thereby further removing the requirement to respond to one's own images). It should be noted that this simple technique, which can be taught by any trained health professional or paraprofessional in a single session and can be supported by simple cassette tape instructions and reading materials, actually combines mental and skeletal muscle relaxation, as well as expectancy and "placebo" effects.

Carrington *(24)* proposes a similar procedure in terms of the basic component processes. However, her procedure differs from Benson's in a number of important respects. Whereas Benson recommends that people practice his technique 15 to 20 minutes in the morning and evening, Carrington recommends that people practice on a more ad lib basis, and that, ideally, they should practice in actual stressful situations. In addition, Carrington encourages persons to select their own "mantras" so as to make the procedure more personally relevant and pleasant. Carrington claims that these changes, plus others, lead to increased adherence. However, there are currently no published studies that carefully compare the different relaxation procedures in regard to their actual clinical efficacy and long-term adherence.

Biofeedback has evolved over the past 10 years as a means of teaching specific voluntary control over particular muscles or visceral responses *(25)*. As shown in Figure 29.1, biofeedback can be seen as the use of electronic sensors to make normally unconscious physiological feedback processes conscious and thereby increase the capacity of the person to exercise self-control. Whether biofeedback training procedures substantially augment the effectiveness of various relaxation procedures used by themselves is controversial *(26)*. It is likely that biofeedback training is, per se, not essential for many patients with stress-related problems. However, research does suggest that biofeedback is important as an adjunct to stress management procedures for at least two major reasons:

1. It helps convince the patient that he can actually voluntarily control his physiological responses, and that psychosocial stress does, in fact, elicit stress responding.
2. It provides reinforcement for the patient and therapist regarding the patient's progress over time.

It allows both patient and therapist to discover what relaxation procedure (or combination of stress management procedures) is or is not effective in reducing the physiological symptoms of stress responding. As Schwartz *(27)*

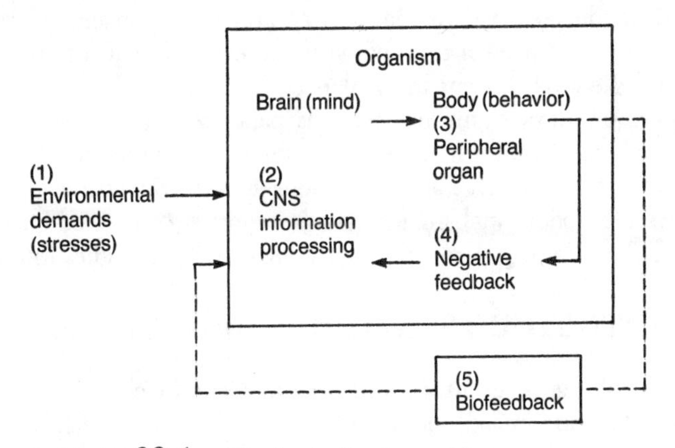

FIGURE 29.1 The biofeedback procedure

Simplified block diagram depicting (1) environmental demands influencing (via exteroceptors, not shown), (2) the brain's regulation of its (3) peripheral organs, and (4) negative feedback from the periphery back to the brain. Disregulation can be initiated at each of these stages. Biofeedback (stage 5) is a parallel feedback loop to stage 4, detecting the activity of the peripheral organ (stage 3) and converting it into environmental demands (stage 1) that can be used by the brain (stage 2) to increase self-regulation *(36)*.

emphasizes in this context, biofeedback should not be used simply as a self-regulation training technique, but rather as a clinical "research" tool essential to effective and responsible clinical practice.

Various mental self-control procedures have been devised for helping patients decrease stress responding. In addition to systematic desensitization, where anxiety provoking imagery is coupled with deep muscle relaxation to reduce stress, specific stress "inoculation" programs have been developed for helping people cope with pain and distress *(28)*. For example, Turk used an experimental pain task to document that stress inoculation training (consisting of imagery-rehearsal role playing and specific mental self-statements) resulted in a 100 percent increase in a subject's ability to endure the pain, whereas morphine alone led to only a 30 to 60 percent increase in a subject's ability to endure the pain *(29)*. It should also be noted that hypnosis and various other suggestion procedures are sometimes used to further enhance such effects.

It is generally accepted that specific combinations of stress management techniques can help certain people reduce their response to stress, and that these effects are not due to simple expectancy or "placebo" factors. However, the research has not advanced to the point that enables clinicians to predict with any precision what kind of person, with what kinds of problems, will respond best to what combinations of procedures. Furthermore, little research has been conducted to date that has combined, for example, relaxation training with assertiveness training with the explicit goals of changing the person and the environment in a balanced fashion.

However, most significant for this paper is that at the time it was written, only two studies had been published that systematically evaluated the use of a stress management technique in a work setting *(30, 31, 32)*. However, a few additional studies have been either completed but not published, or they are in progress. A brief review of these studies follows.

WORK SITE STUDIES OF STRESS MANAGEMENT

Peters and colleagues recently published a major study comparing the effects of daily "relaxation breaks" on five self-reported measures of health, performance, and well-being *(30)* and measures of systolic and diastolic blood pressure and heart rate *(31)*. The work was done at the Converse Rubber Company, a subsidiary of the Eltra Corporation. For 13 weeks, 126 volunteers filled out daily records and reported for biweekly blood pressure measures. After four weeks of baseline monitoring, they were divided randomly into three groups: Those in Group A were taught Benson's technique for producing the relaxation response, those in group B were instructed to sit quietly and relax any way they wanted, and group C received no instructions. Groups A and B were asked to take two 15-minute relaxation breaks daily.

After an eight-week experimental period, the greatest mean improvements on every index occurred in group A, the least improvements occurred in group C, and group B was intermediate. Differences between the mean changes in group A versus group C reached statistical significance ($P < .05$) on four of the five indices: symptoms, illness days, performance, and sociability-satisfaction. The relationship between amount of change and rate of practicing the relaxation response was different for the various indices. Although fewer than three practice periods per week produced little change on any index, two daily sessions appeared to be more than necessary for many persons to achieve positive changes. Interestingly, somatic symptoms and performance responded with less practice of the relaxation response than did behavioral symptoms and measures of well-being.

The results for blood pressure paralleled the self-report measures. Although the subjects generally were normotensive, the decreases in systolic blood pressure from the beginning to the end of the study were −11.6 mm Hg for group A, −6.5 mm Hg for group B, and +0.4 mm Hg for group C; mean diastolic blood pressures decreased by −7.9, −3.1, and −0.3. Moreover, within group A, the higher the initial blood pressure, the greater the decrease with relaxation training.

An interesting, serendipitous finding occurred for the blood pressure data. Both systolic and diastolic blood pressures rose in all three groups for session 6 and fell again for session 7 in groups A and B but not group C. The project was apparently initiated shortly before the company experienced the effects of a nationwide economic recession. As a result, the company initiated a series of layoffs, most of which were implemented on three consecutive Fridays, beginning at session 5 and ending at session 6. Although only 5 percent of the total corporate staff were laid off, another 10 percent were offered and accepted other positions in the company. During sessions 5 and 6, many participants mentioned their anxiety about their job security or increased workload or their concern for colleagues who had been forced to move or leave. Peters and co-workers *(30)* hypothesize that stress mounted over those weeks, and blood pressure increased accordingly. Other investigators *(33)* have reported such increases in blood pressure during the anticipation phase of factory shutdowns among employees who eventually lost their jobs. Peters and co-workers *(29)* also offer the hypothesis that blood pressures may have returned to the original levels more rapidly in groups A and B than in C (session 7) because of some effect of the relaxation practice.

The preceding data are clearly encouraging and beg to be replicated and extended under more controlled conditions. Peters and co-workers *(29)* point out some of the limitations of their experiment, including the lack of control for positive expectancy effects, the lack of follow-up data, the lack of actual data regarding subsequent use of health services, and so forth. Furthermore, their study was not performed on patients who were seeking help for specific stress-related disorders. On the other hand, the data illustrate the potential of incorporating into an industrial setting a relatively simple relaxation procedure that could have beneficial effects on health and work performance. Peters and co-workers note that if the relaxation response proves capable of maintaining lowered blood pressure in normotensive subjects, "it might become a most useful component of preventive as well as therapeutic programs."

Regarding cost, Peters and co-workers comment: "the relaxation response is particularly attractive as a preventive measure since it costs only the time involved to practice, has no known side effects, and is reported to be a pleasant and personally rewarding experience by those who elicit it regularly" *(29)*. However, precise cost-benefit ratios have not been computed.

Carrington and colleagues *(32)* have recently compared Benson's relaxation technique with Carrington's meditation technique and with Jacobson's progressive muscle relaxation technique, all in an industrial setting. A total of 154 New York Telephone employees, self-selected for stress, were taught one of the three techniques. At five and a half months the treatment groups reported clinical improvements in symptoms of stress, but only two of the techniques (Carrington's and Benson's) showed significantly more symptom reduction than a group of control subjects. Carrington and colleagues *(32)* note that both Carrington's and Benson's techniques are quite cognitive in nature, whereas Jacobson's technique is more directly somatic. Interestingly, the more cognitive techniques resulted in high compliance after five and a half months (78 percent), suggesting that they are relatively easy to learn and have some subjectively perceived benefits. However, since no physiological data were collected in the study, it is not known whether the physical health of these subjects improved. Nonetheless, like the findings of Peters and colleagues in the studies cited above, those of Carrington and co-workers are encouraging and should be followed up with more comprehensive and controlled investigations.

In the cited studies by Peters and co-workers, the relaxation training was purely voluntary. The company did not reinforce the employees for learning stress management by giving them, for example, time off from work to learn and practice the skills. However, an excellent model of a corporation-supported, comprehensive approach to stress management was developed by Manuso and colleagues for Equitable Life Assurance Society of the United States. They have established an Emotional Health Program, staffed with a clinical psychologist, a psychiatrically oriented physician, a clinical psychology intern, and a counselor. Liaisons are maintained with outside mental health practitioners, universities, and hospitals. The program is more than just a referral service. It is concerned with the detection, prevention, education, treatment, referral, and follow-up of troubled employees. All services are free and on company time, along with all other medical services. The

Emotional Health Program is physically housed in the Employee Health Services Department, thereby enabling the delivery of multimodality (psychological and medical) services. All services are confidential, ensuring that the workers can freely pursue and therefore, it is hoped, resolve their problems.

Manuso has just completed his first study examining both the clinical and the cost effectiveness of providing biofeedback and other stress management procedures in the treatment of 15 subjects with headaches and 15 subjects with general anxiety. These subjects held a wide variety of job responsibilities, from filing to upper managerial jobs. Their average annual salary was $14,000. The subjects were included in the study if their average symptom activity and their symptom's history met a minimum standard; namely, if symptoms had been present for at least 5 years and if the average symptom activity for a two-week baseline period (assessed by using a daily log procedure) was "moderate" to "severe." A unique feature of this study was that an additional and different subject population of 30 was used to report on the extent to which significant others at the worksite with stress-related symptoms interfered with the respondent's ability to work. From their percentage estimates, a quantitative effect of interference could be generated. Bosses, closest co-workers, and subordinates were used as significant others at the worksite.

The experiment consisted of three phases: the pretreatment baseline phase, during which two no-feedback, electromyographic (EMG) measurement-only sessions were administered; the treatment phase, consisting of five weeks of frontalis EMG biofeedback training with three sessions per week (the average number of treatment sessions per subject was 13), and the posttreatment follow-up phase, which consisted of two no-feedback, EMG measurement-only sessions over a two-week period, taking place three months following the last treatment phase session. Subjects were grouped according to primary diagnosis (headache versus anxiety) and they served as their own controls.

Numerous before-and-after measures were taken in this study, including personality scales, health center medical records, daily logs (including medicine used), total interference and work interference due to the presence of symptoms, and so forth. The stress management training included muscle relaxation, breathing exercises, and imagery techniques, as well as the biofeedback training. Since a complex stress-management package was used, it is not possible to attribute the results to any one component or combination of components in the treatment package.

The results were striking because, on the average, improvement occurred in every measure taken. Both groups showed statistically significant decreases in symptoms and increases in work-related satisfaction and effectiveness. Importantly, both anxiety and headache subjects significantly decreased their visits to the health center for both stress-related and other complaints from the period three months before treatment (5.75 visits per subject) to three months after treatment (1.70 visits per subject). Whereas during the three pretreatment months, all subjects had visited the health center because of stress-related symptoms, only five anxiety subjects and two headache subjects visited the health center during the three posttreatment months.

These initial results, although striking, must be viewed with caution. There were no control groups in the study. Therefore, one cannot conclude with certainty that similar results would not have been obtained if the subjects had been, for example, placed on drugs (or a "new" drug with potential placebo effects). Furthermore, the follow-up period is short. On the other hand, only subjects who had long-standing symptoms were selected, and all subjects were currently being seen for traditional biomedical treatment. This argues in favor of the interpretation that sizable decreases in headaches and anxiety observed in these patients were due, at least in part, to the comprehensive stress management program.

Manuso is careful to consider the cost-benefit aspects of this work.

> The estimated average annual pretreatment corporate costs of employing one person with chronic anxiety or headache was found to be $3,394.50. The costs to the corporation of an experimental subject-employee are considerable though not patently obvious. The costs involved four factors: namely, visits to the Employee's Health Center ($473.14), time away from the job in order to visit the Health Center ($56.61), work interference due to symptoms ($2,206.95), and metainterference affecting bosses ($72.80), co-workers ($542.88), and subordinates ($42.12). These costs were present even though subjects were receiving traditional medical treatment, involving diagnosis, prescription of appropriate medications, and follow-up by a physician. It will be noted that absenteeism figures are conspicuously absent from this accounting. This is because, on the average, Anxiety and Headache subjects were absent only 4.27 days per year, which is not significantly different from the overall corporate average of 3.92 (two-tail Z = 1.00).
>
> The estimated average annual post-treatment corporate costs of employing one person with chronic anxiety or headache were found to amount to $532.68. These costs, when compared to the corresponding pretreatment costs, indicate average savings of $2,861.82 annually per subject. By extrapolation, the expected 3-year savings to the corporation afforded by reduced symptom activity for all 30 subjects would amount to $202,945.05 minus the costs treatment of $24,622.50, which equals $178,322.55. Whereas earlier work (Manuso, 1978) indicated a 1:3 cost-benefit ratio, taking into account a 23% dropout rate, the present research suggests that the ratio averages 1:5.52 for each of the first three years following treatment. This represents a considerable return on investment. *(34)*

These figures must be viewed as tentative, and they are probably overly optimistic. They do, however, illustrate the potential for stress management programs to have some cost as well as clinical effectiveness. They also point to the need for more comprehensive clinical studies to be conducted in the future.

SUGGESTIONS FOR THE FUTURE

With our current knowledge, it is appropriate to consider incorporating various stress management techniques into occupational settings on an experimental basis. Despite the promise of present findings, much needs to be learned through future research that is relevant to industry, labor, and science. Although it is tempting to simply bring some stress management consultants into an industrial setting to conduct a program or two, this approach will not prove valuable in the long run. What is needed are clinical research studies in which relevant health and work variables are measured concurrently, with appropriate short- and long-term evaluation included as a necessary component of the program. Evaluation should not be viewed as necessary only for the initial developmental stages of such programs. Rather, evaluation (both clinical and cost effectiveness) should be incorporated as a standard component of such programs if future research proves them to be valuable in occupational settings.

Clearly, industry needs to consider how it can promote stress management (as well as health more broadly) by changing its incentive structure. In fact, it may even prove cost-effective to reinforce workers to take "relaxation breaks" *(30, 31)*, which may take a variety of forms (including mental and physical relaxation exercises, plus other recreational-relaxing activities). It may prove cost-effective also to change the work setting to better match the needs and physiological limits of the people working in the settings. It is probable that labor will resist simple stress management programs if these programs are offered in the absence of other needed changes in work settings. Requiring the worker to continue to cope with more and more job stress is not a final solution. At some point the strain will become too great, and everyone (both management and labor) will suffer the consequences. Industry must strive toward reaching a balance between the requirements of the work setting and the worker's capacity to meet those requirements. Industry could make a major contribution to society at large by providing an incentive for health behavior.

As described in the "Models Linking Stress to Illness," stress management should not be isolated. Stress management training can have positive spinoffs, such as reducing drug usage, improving diet, and promoting exercise.

MODELS LINKING STRESS TO ILLNESS

Numerous models link stress to illness. Furthermore, confusion and inconsistency exist even in the use of the term "stress" *(35)*. A major source of confusion is whether the term should be used to refer to (*a*) a stimulus in the environment (for example, the loss of one's job), (*b*) the interaction between the person and the environment (for example, how the person perceives the loss of the job), or (*c*) the response of the person (for example, increased blood pressure, circulating catecholamines, psychological depression) to the environment. Selye *(12)*, who pioneered the concept of the General Adaptation Syndrome, used the term "stress" to refer to a general stress response of the person, defined stimuli that caused "stress" as "stressors."

In physics and engineering, however, the term "stress" is used to refer to the stimulus in the environment. The term "strain" is used to refer to the person's response to stress. By these definitions, stress management would refer only to modify the external stresses, while strain management would refer to actually modifying the person's response to the external stresses.

In this paper, I use the term "stress" to refer to the environmental stimulus; the term "distress" to refer to the person's perceived negative reaction to the environmental stresses, and the term "stress response" (or strain) to refer to the physiological and behavioral consequences of stress. However, I use the term "stress management" in the broadest sense to refer to changing any aspect of the environment or person that will decrease stress response (strain) and promote health.

Figure 29-1 (page 239) shows a simplified, useful means I developed *(36)* for summarizing various models that link stress with illness. Stage 1 (environmental demands) refers to any environmental stress that can potentially place strain on any part of the brain (stage 2) or body (stages 3 and 4). Note that a general stress model of disease includes "simple" physical, chemical, or biological stimuli (temperature, pollutants, or germs), as well as more "complex" psychosocial stimuli such as the demands of being an air traffic controller, as potential stresses than can place strain on the brain or body. A general stress model is useful because it classifies psychosocial stresses as one subgroup of all potential stresses, and therefore views psychological and biological stimuli within a common, biobehavioral framework.

Whereas simple physical-chemical-biological stresses can directly place strain on the body (stages 3 and 4) without necessarily involving the central nervous system (the brain, stage 2), psychosocial stresses operate only on the body (stages 3 and 4) indirectly via the central nervous system (the brain, stage 2). The brain stores all the person's past experiences and therefore modifies the ultimate physiological or behavioral response (stage 3) to the stage 1 environmental demand.

Figure 29-1 illustrates, then, an important point regarding the effects of psychosocial stress (stage 1) on the body (stage 3). It follows that strain on the body (stage 3) is always a complex interaction of (*a*) the nature of the environmental stress (stage 1), (*b*) the way the person perceives the stress and reacts to it (stage 2), (*c*) the sensitivity of the body (stage 3) to neural and humoral control from the brain (stage 2), and (*d*) feedback from the body (stage 4) back to the brain regarding the degree of strain on the organs (stage 3) and the brain's (stage 2) response to the feedback.

Genetics, nutrition, diet, exercise, previous disease, and so forth can influence stages 2–4, and therefore can modify the person's response to a psychosocial stress. Vulnerability to psychosocial stress can be mediated in part by circulating drugs (for example, from cigarettes or coffee), circulating hormones (for example, during the menstrual cycle), and so on. Nutrition, drugs, diet, and genetics may influence the brain's response to the psychosocial stimuli as well as the organ's sensitivity to neural and humoral responses from the brain. The point to be emphasized is that psychosocial stress (stage 1) never acts on the body (stage 3) in a vacuum, but rather it involves a complex interaction of biological and psychological processes that mediate the stress response.

This analysis of stress is useful for other reasons. It illustrates that there are various mechanisms by which psychosocial stress can increase susceptibility to infectious disease as well as influence healing of all diseases. As originally postulated by Selye *(12)*, it is now known that the immune system is modulated, in part, by the brain (stage 2). If the immune system does not function properly, this will increase a person's susceptibility to all kinds of physical (and genetic) disease, as well as recovery from illness. Using the terminology shown in the diagram, stage 1 psychosocial stresses can, via stage 2, disrupt the immune system in the body (stage 3) such that other stage 1 physical, chemical, or biological stresses can more easily act directly on the body (stage 3) to cause disease.

As more is learned about the central role that the brain plays in the expression of (*a*) psychological processes *(11)* and (*b*) physiological regulation, it becomes clearer how psychosocial factors can play a role in the pathogenesis, treatment, and recovery of all disease. Hence, it is understandable why researchers such as Engel *(37)* are calling for the development of new medical models that take a more integrated, "biopsychosocial" approach to health and illness.

There are numerous other implications of the preceding structural analysis of stress and illness. For example, it becomes clear how psychosocial stresses may modulate the brain in such a way as to lead the person to (*a*) change his or her diet to possibly reduce symptoms of distress (from stage 4), (*b*) take drugs such as alcohol to deaden the experience of distress, (*c*) become depressed, have difficulty sleeping, and therefore not get enough exercise, and so forth. It is well known that psychosocial stress can disrupt healthful behavior, which in turn contributes to disease. In other words, for some individuals, psychosocial stress may be an important mediating factor in the maladaptive behavior. Stress management programs can sometimes have beneficial spinoff effects of reducing people's maladaptive needs for food and drugs, increasing energy, and spurring the desire to exercise, all of which in turn help to promote health.

Another example concerns Cannon's *(38)* concept of homeostasis and its relationship to disease. Cannon argued that the body is designed to maintain physiological levels within certain limits despite demands placed on the body by external physical, chemical, biological, or psychosocial stresses. Figure 29-1 illustrates how homeostasis works. Feedback (stage 4) from the body (stage 3) is processed by the brain (stage 2) in such a way as to readjust the regulation of the organ in question (stage 3) so as to maintain certain healthful limits. Much of this self-regulatory system is unconscious and appears involuntary. However, symptoms of distress (that is, pain) may emerge from the body (stage 4). The purpose of such pain stimuli is to lead the person (via the brain, stage 2) to (*a*) modify the source of stress in the environment (stage 1), (*b*) leave the environment for the sake of the organ's health (stage 2 leaving stage 1), (*c*) modify the person's reaction to the external source of the stress (by learning how to relax), (*d*) repair the injured organ (direct modification of stage 3), or (*e*) simply remove the pain per se (achieved by modifying stages 4 or 2 via surgery, drugs, or psychology).

The concept of the need to "treat the cause rather than the symptom" can be restated as the need to modify or eliminate the stress (stage 1) rather than simply eliminate the symptoms of distress (stages 4 or 2). It should be noted that simply repairing the organ (stage 3) leaves the psychosocial stresses intact (stage 1), so that other problems may develop in the future. Furthermore, simply eliminating the distress (via stages 4 and 2) without also affecting the environment (stage 1) results in removing the very mechanism biologically designed to protect organisms from dangerous environments in the first place. Removing these protective feedback loops can, in my terms, be "disregulatory," since it allows psychosocial stresses to increase rather than keeping them in balance *(36)*.

I raise these issues to illustrate both the complexity of the problem linking stress to illness as well as the potential for improvement. As more is learned about the role of the brain in mediating responses to psychosocial stress, the more we will be able to understand the extent to which psychosocial stress can contribute to disease, the more we can consider modifying the person's perceptions and reactions by using behavioral procedures to minimize the effects of psychosocial stress, and the more we can appreciate the need to take an integrated approach to stress management. Industry and labor can work together to both minimize sources of stress in the work environment (stage 1) as well as develop better means of coping with the work environment (stage 2) for the sake of the health of the worker and industry as a whole.

Moreover, health programs aimed specifically at changing drug usage (including cigarettes and alcohol), diet, and exercise can have positive spinoffs by helping persons cope with stress. As noted by Benson *(22)*, Carrington *(24)*, and Meichenbaum *(28)*, stress management is a skill useful to any person. The work setting is but one setting, albeit a significant one, where such skills are of value. It should also be recognized that stress management need not be viewed only as a means of preventing illness, but also as a means of promoting health. Many of the relaxation and cognitive exercises are inherently pleasant and bring other personal rewards, as does regular exercise.

As noted by Manuso *(9)*, one way that industry may be able to promote the development of stress management programs is to offer predoctoral or postdoctoral fellowships in clinical and health psychology and related disciplines. Most clinical psychologists, for example, do not have experience in occupational settings. To encourage psychologists and other health professionals to apply their skills to problems relevant to occupational settings, a training-incentive program should be established. It should be recognized that developing such internship programs is also cost-effective in that interns typically work more hours for less pay as part of the training experience. Nurses and physicians also can be trained to administer some types of stress management programs, and this option too should be pursued.

It is not possible to present in this paper detailed suggestions regarding specific directions for future research and applications, including possible alternative structures for incorporating stress management programs into occupational settings. However, I have provided a general introduction to the problems and promise of stress management as applied to occupational settings. The challenge is becoming clear. Whether the challenge will be met depends on the cooperation and collaboration of industry, labor, government, and the behavioral and biomedical sciences in the context of the emerging field of behavioral medicine *(7, 8)*.

REFERENCES

1. Selye, H. *Stress in health and disease.* Woburn, Mass.: Butterworth Publishers, 1976.
2. Weiner, H. *Psychobiology and human disease.* American Elsevier Co., New York, 1977.
3. Williams, R. B., and Gentry, W. D., eds. *Behavioral approaches to medical treatment,* Cambridge, Mass.: Ballinger Publishing Co., 1977.
4. McLean, A. ed. *Occupational stress.* Springfield, Ill.: Charles C. Thomas, 1974.

5. McLean, A. ed. *Reducing occupational stress.* Proceedings of a conference. DHEW Publication no. (NIOSH) 78–140. Washington, D.C.: Government Printing Office, 1977.
6. McLean, A. *How to reduce occupational stress.* Reading, Mass.: Addison-Wesley Publishing Co., 1979.
7. Schwartz, G. E., and Weiss, S. M. *Proceedings of the Yale Conference on Behavioral Medicine.* DHEW Publication no. (NIH) 78–1424. Washington, D.C.: Government Printing Office, 1978.
8. Schwartz, G. E., and Weiss, S. M. Behavioral medicine revisited: An amended definition. *Journal of Behavioral Medicine* 1:249–252; (a) 250 (1978).
9. Manuso, J. Psychological services and health enhancement: A corporate model. In *Linking health and mental health: Coordinating care in the community.* Sage Annual Review of Mental Health, vol. 2. Beverly Hills: Sage Publications, 1981.
10. Chesney, M. A., and Feuerstein, M. Behavioral medicine in the occupational setting. In *Behavioral medicine,* edited by J. McNamara. Kalamazoo, Mich.: Behaviordelia, 1979.
11. Weiman, C. G. A study of occupational stressor and the incidence of disease/risk. *Journal of Occupational Medicine* 19:119–122 (1977).
12. Selye, H. *The stress of life.* New York, McGraw-Hill, 1956.
13. Zorn, E. W.; Harrington, J. M.; and Goethe, H. Ischemic heart disease and work stress in West German sea pilots. *Journal of Occupational Medicine* 19:762–765 (1977).
14. Dutton, L. M., et al. Stress level of ambulance paramedics and fire fighters. *Journal of Occupational Medicine* 20:111–115 (1978).
15. Holmes, T. H., and Rahe, R. H., The social readjustment scale. *Journal of Psychosomatic Research* 11:213–218 (1967).
16. Chadwick, J. H.; Chesney, M. A; and Feuerstein, M. *Psychological job stress and coronary heart disease.* Cincinnati: National Institute of Occupational Safety and Health, in press.
17. Moos, R. H., Insel, P. M., and Humphrey, B. *Work environment scale.* Consulting Psychologists, Palo Alto, Calif., 1974.
18. Eysenck, H. J., and Eysenck, S. B. G. *Eysenck personality inventory.* San Diego, Calif.: Education and Testing Service, 1968.
19. Novaco, R. *Anger control: The development and evaluation of an experimental treatment.* Lexington, Mass.: D. C. Heath and Co., 1975.
20. Jacobson, E. *Progressive relaxation.* Chicago: University of Chicago Press, 1938.
21. Schultz, J. H., and Luthe, W. *Autogenic training: A psychophysiological approach in psychotherapy.* New York: Grune and Stratton, 1969.
22. Benson, H. *The relaxation response.* New York: William Morrow and Co., 1975.
23. Goleman, D. J., and Schwartz, G. E. Mediation as an intervention in stress reactivity. *Journal of Consulting and Clinical Psychology* 44:456–466 (1976).
24. Carrington, P. *Freedom in meditation.* Garden City, N.Y.: Doubleday and Co., 1977.
25. Schwartz, G. E., and Beatty, J., eds. *Biofeedback: Theory and research.* New York: Academic Press, 1977.
26. Silver, B. V., and Blanchard, E. B. Biofeedback and relaxation training in the

treatment of psychophysiological disorders: Or are the machines really necessary? *Journal of Behavioral Medicine* 1:217–239 (1978).

27. Schwartz, G. E. Research and feedback in clinical practice: A commentary on responsible biofeedback therapy. In *Biofeedback: Principles and practices for clinicians,* edited by J. Basmajian. Baltimore: William and Wilkins Co., 1978.
28. Meichenbaum, D. *Cognitive-behavior modification.* New York: Plenum Publishing Corp., 1977.
29. Turk, D. An expanded skills training approach for the treatment of experimentally induced pain. Doctoral dissertation, University of Waterloo, Waterloo, Ontario, 1976.
30. Peters, R. K.; Benson, H.; and Porter, D. Daily relaxation response breaks in a working population: I. Effects on self-reported measures of health, performance and well-being. *American Journal of Public Health* 67:946–953 (1977).
31. Peters, R. K.; Benson, H.; and Peters, J. M. Daily relaxation response breaks in a working population: II. Effects on blood pressure. *American Journal of Public Health* 67:954–959 (1977).
32. Carrington, P., et al. The use of meditation: Relaxation techniques for the management of stress in a working population. *Journal of Occupational Medicine* 22:221–231 (1980).
33. Kasl, S. V., and Cobbs, S. Blood pressure changes in men undergoing job loss: A preliminary report. *Psychosomatic Medicine* 32:19–39 (1970).
34. Manuso, J. Corporate mental health programs and policies. In *Strategies for public health,* edited by L. Ng and D. Davis. New York: Van Nostrand Reinhold Co., 1980.
35. Mason, J. A historical view of the stress field. *Journal of Human Stress* 1:6–12 (1975).
36. Schwartz, G. E. Psychosomatic disorders and biofeedback: A psychobiological model disregulation. In *Psychopathology: Experimental models,* edited by J. D. Maser and M. E. P. Seligman. San Francisco: W. H. Freeman and Co., 1977.
37. Engel, G. L. The need for a new medical model: A challenge for biomedicine. *Science* 196:129–136 (1977).
38. Cannon, W. B. *The wisdom of the body.* New York: W. W. Norton and Co., 1932.

30

The Physical Activity Component of Health Promotion in Occupational Settings

William L. Haskell, Ph.D., and Steven N. Blair, P.E.D.

Most adults believe that regular exercise is important for good health, and many also state that their own health would benefit from more exercise *(1)*. This positive attitude by the general public toward the potential health benefits of exercise is consistent with, and probably substantially influenced by, the admonitions of numerous health organizations that regular exercise is important for prevention of chronic disease and rehabilitation *(2, 3)*. Available data, however, indicate that a substantial discrepancy exists between the general public's attitudes or beliefs about the health benefits of exercise and their actual exercise habits. Without adjusting for the likelihood of substantial overreporting of actual activity performed, fewer than 30 percent of American adults probably meet the American Medical Association and American Heart Association recommendation of "a good bout of exercise at least three times per week" *(1, 4)*.

This article was first published in *Public Health Reports* (March–April 1980), pages 109–118, and appears here in slightly revised form.

Dr. Haskell is with the Stanford University School of Medicine, and Dr. Blair is with the University of South Carolina School of Public Health.

Continued expansion of the physical activity boom that started in the mid-1970s could result in a substantial change in the exercise profiles of American adults by the early 1980s. Can members of the health profession take advantage of this new interest in vigorous exercise (and associated changes toward a more healthful life-style)—demonstrated by a modest but growing segment of the population—and extend it to a sufficiently large proportion of sedentary, working adults so that a measurable change in health status and productivity will result?

To evaluate the potential role of physical activity in employer-employee health promotion programs, our focus here is on the following areas: (*a*) scientific evidence of health and job-related benefits resulting from increased physical activity, (*b*) factors contributing to the successful initiation and maintenance of physical activity by adults, and (*c*) implementation of exercise programs in industrial settings.

HEALTH AND JOB-RELATED BENEFITS OF EXERCISE

The belief that regular exercise provides significant health benefits has existed since the time of Hippocrates. Currently, there are both vocal opponents and advocates of the health potential of exercise. Their positions on the subject range from the belief that vigorous exercise is a panacea for all that ails man to the belief that it provides no health benefits and is a hazard to be avoided by every adult. It is evident that truth is to be found somewhere between these two extremes. For selected chronic disorders highly prevalent among the employable portion of our population, exercise can have a unique or synergistic beneficial effect when incorporated into a health-promoting life-style. As with many other changes in health-related habits, the scientific evidence that demonstrates a cause-and-effect relationship between a change in the habit and a reduction in the frequency or severity of various health problems still is not definitive and, in some cases, still merely circumstantial. On the other hand, there is substantial evidence that for some health- and job-related conditions, an appropriate increase in the proper exercise by previously inactive adults can produce direct beneficial effects.

HEALTH BENEFITS OF INCREASED EXERCISE

Evidence from a variety of sources supports the belief that an increase in habitual physical activity promotes better health. The degree of conviction regarding the relationship between exercise and a specific benefit should depend on the scientific rigor of the studies performed to support the relationship and the reproducibility of the results. The utility of the results depends on their generalizability to a reasonable proportion of the general population if a change in exercise is to contribute significantly to an employee health promotion program. Factors influencing generalizability include the characteristics of the exercise program (type, intensity, amount) and of the study population (age, health status, and so on). A brief review of health benefits ascribed to regular exercise that have some scientific documentation follows.

Exercise and Coronary Heart Disease

The greatest impetus for considering exercise a health modality is its potential role in prevention of heart diseases and rehabilitation. Substantial evidence derived from observational studies demonstrates an association between an active life-style on the job or during leisure time and a reduced risk of coronary heart disease morbidity and mortality. More active persons tend to have fewer heart attacks; when they do occur, such persons are older, and the attacks seem to be less severe *(5, 6)*. This benefit appears to increase with increasing amounts of exercise and is, at least in part, independent of other established risk factors *(7, 8)*. As with every other heart disease risk factor (blood pressure, cigarette smoking, cholesterol, and so forth), there still is no definitive evidence that an increase in exercise by previously sedentary persons will result in reduced heart disease morbidity and mortality. It is highly unlikely that an adequate study to answer this question will be completed during the next decade; thus any decision about the role of exercise in heart disease prevention must be based on less than definitive data.

Exercise may reduce the risk of heart attack by means of favorable alterations of several biochemical or circulatory functions. Some major changes known to occur from exercise that may contribute to a reduction in coronary heart disease manifestations include (*a*) alterations in fat and carbohydrate metabolism resulting in a more favorable blood lipoprotein profile *(9)* and thus potentially contributing to a reduced rate of coronary atherosclerosis *(10)*; (*b*) a reduction in sympathetic nervous system activity *(11)*, resulting in a lower workload of the heart through a decrease in heart rate and, in some persons, a decrease in blood pressure *(12)*; and (*c*) possibly enhanced electrical stability of the myocardium *(13)*. Current evidence does not support the notion that exercise produces a significant increase in the coronary vascular bed as a result of either enhanced coronary artery size or the development of collateral vessels *(14)*.

Despite the numerous limitations in the studies supporting the potential of exercise in preventing heart disease, it seems justifiable to include exercise as a component of multifactor risk reduction programs. Exercise can help control other established risk factors, it can be implemented at a relatively low cost by a large portion of the employed population, and it also may provide other health- and job-related benefits.

Exercise and Weight Control

It is generally accepted that excess body weight is associated with increased morbidity and mortality. Changes in body composition (percentage of total weight that is body fat) in adults depend primarily on the balance between calorie intake (food) and calorie expenditure (exercise). Modest amounts of exercise repeated frequently throughout the day can significantly alter a person's energy balance and shift it from positive to negative at no cost for special equipment or facilities. With increasing age there tends to be a small but significant decrease in resting metabolic rate associated with a decrease in muscle mass. Much of this decrease in metabolic rate can be prevented by regular activity of sufficient intensity to retain muscle mass. Even a 5 percent decrease in metabolic rate with aging can result in the potential for gaining 8 to 10 pounds of body fat per year at the same calorie intake. An increase in habitual activity should be considered a component of all weight maintenance or control programs.

Exercise and Psychological Status

Much of the research on exercise and its effects on psychological status— especially tension, anxiety, and depression— supports the conclusion that exercise that results in improved physical fitness has psychological benefits for many adults. Exercise has been associated with an improved sense of well-being *(15)*, better sleep patterns *(16)*, and reduced muscle tension *(17)*; it also has been correlated with objective demonstrations of reduced anxiety, depression, and hostility *(18)*. Improvement in psychological parameters in relation to exercise appears to be greatest in persons initially most unfit physically and psychologically *(19)*. The physiological bases for such changes in psychological status have not been clarified, but the known changes in central nervous system activity and reduced circulating catecholamine levels may be a significant factor. Some researchers have assumed that the experience of exercising in a group could provide social-psychological benefits that would assist in stress management, and that an improved self-image might bolster the resources needed to deal with stress and tensions. Moreover, just the escape from routine job-related tasks to exercise alone or in a social setting, with or without competition, may provide a natural and socially acceptable release from stress or tension.

Howard and colleagues *(20)* studied the coping techniques used by mid-level managers to handle job-related stress. Participants were asked which of 10 coping techniques they used. They also completed a stress symptom checklist and underwent medical and behavioral evaluations. Those who engaged in physical exercise as a coping technique had fewer stress symptoms than those who did not. This technique appeared to be the third best (of 10) in preventing stress symptoms. These authors concluded that the five best coping techniques were (*a*) building resistance by regular sleep, exercise, and good health habits, (*b*) compartmentalizing work and nonwork activity, (*c*) engaging in physical activity, (*d*) talking with peers on the job, and (*e*) withdrawing physically from the stress situation.

Exercise and Orthopedic Limitations

Inappropriate or unaccustomed exercise, especially running, jumping, or active sports, can cause orthopedic injuries that can be uncomfortable or debilitating, while a progressive exercise program properly implemented can aid in the rehabilitation of selected orthopedic problems and contribute to the prevention of others. Strength exercises to enhance abdominal muscle tone and stretching exercises to maintain lower back flexibility significantly reduce the frequency of low back discomfort and disability *(21)*. Whether such exercise decreases the likelihood of orthopedic injury during falls has not been firmly established, but increased muscular strength, greater joint flexibility, and enhanced bone density as a result of proper exercise by sedentary adults may provide some protection.

Other Health Benefits

Because of the numerous biochemical changes that occur during vigorous exercise of large muscles, a variety of health benefits have been ascribed to exercises; however, these benefits are rarely documented scientifically. It has been observed, for example, that exercise decreases blood sugar levels and reduces the need for medication by some diabetic patients and that the arterial blood pressure levels at rest and submaximal exercise are reduced in some normal and hypertensive patients, but there is no evidence that exercise prevents either diabetes or essential hypertension. And the notion that exercise may enhance gastrointestinal or sexual functions is speculation, based on subjective observations or wishful thinking. Only further research will provide the answers to these and a variety of other questions that relate exercise to improved health and general well-being. Available data, however, support the inclusion of regular exercise in a health promotion program.

Exercise and Job Performance

A major benefit of regular physical activity is an increase in physical working capacity (PWC) or physical fitness. PWC is the maximum amount of energy that a person can expend for a short time, and it is expressed in relation to the energy used by a person at rest. If the energy expended at rest is 1 metabolic unit (MET), then 2 METS represent double resting energy expenditure and 10 METS would indicate a 10-fold increase in energy expenditure. With this multiple of the resting energy concept, PWC can be expressed in relative terms: that is, 5 METS is the same relative increase in metabolism for a person regardless of size, age, or sex. Thus, the relative physiological stress of a given physical task can be expressed as the PWC-MET ratio. For example, a task requiring 4 METS represents 50 pecent of the capacity of a person who has a PWC of 8 METS, but only 30 percent of the capacity of a person with a PWC of 12 METS.

Since people can work for extended periods (eight hours or more) at no more than 20 to 25 percent of their capacity, this PWC-MET ratio is significant. Many sedentary U.S. adults have PWCs of 8 to 10 METS; thus their sustained working capacity is only 1.6 to 2.0 METS. Standing costs 1.5 METS and using a riding lawnmower costs 2 METS; therefore many people clearly lack the PWC to engage in even a moderately active life-style.

The implication of these foregoing principles is that even people who have moderately active jobs and leisure-time life-styles need to meet certain PWC standards if they are to avoid chronic fatigue. A reasonable hypothesis might be that an improvement in PWC will result in less general fatigue and a more positive approach to work and, perhaps, life in general.

Much speculation exists about the potential benefit of regular exercise on the job performance of employees in sedentary occupations. Objective documentation of increased productivity as a result of participation in an exercise program has not been performed adequately in the United States Numerous studies conducted in Germany, Eastern Europe, and the U.S.S. have been reported that supposedly demonstrate enhanced job perform of factory or clerical workers as a result of short exercise breaks durin day. Frequently cited are increased productivity and decreased absen sm associated with increased physical fitness and less job boredom and ntal fatigue. When these claims are investigated it appears that they based on conjecture rather than data. We can find relatively little in th cientific literature on the subject. We conclude that controlled studies a necessary before the issue can be resolved.

Durbeck and colleagues *(22)* studied the effects of a 12- nth exercise training program on three heart disease risk factors, exercise apacity, health attitudes, and job attitudes of 237 male employees of th National Aeronautics and Space Administration headquarters in Washington, D.C. Participants in the employer-sponsored exercise program reported that they could work harder than before—mentally and physically—enjoyed their jobs more, and found their normal work routines less boring. Heinzelmann and Bagley, using the same questionnaire as Durbeck *(23)*, reported similar changes in men who participated in an 18-month jogging-type exercise program designed to study the effects of exercise on heart disease risk factors. This study included 239 men from several communities working in a variety of jobs who were randomly assigned to the exercise group and 142 men randomly assigned to a control group. Significantly more men in the exercise group reported increased work performance (60 percent versus 3 percent) and a more positive attitude toward work (40 percent versus 1 percent). Such self-reports of increased work performance as a result of exercise sessions on personal or company time have been reported by others *(24, 25)*, but no study has been conducted in the United States that objectively demonstrates increased productivity as a result of exercise program participation.

Some investigators in North America have observed a decrease in absenteeism among employees who participated in an exercise program. Bjurstrom and Alexiou *(26)* studied the effects of a five-year heart disease prevention program, primarily an exercise training class, on job performance of 847 men and women employed at the New York State Education Department. Of the 99 participants who completed the first year of the program, 55 percent charged less sick leave during their first year of the exercise program than during the preceding year (36.1 hours versus 66.5 hours). A net reduction of 4.7 hours per employee per year was observed when sick leave data for all participating employees were compared for the control and the program years. Although the findings of this and similar studies *(27)* tend to support the ncept that a more physically fit employee will take less sick leave, none the studies had the scientific rigor needed to show a cause-and-effect ionship. The possible self-selection of more healthy employees into the ar se groups and the lack of an appropriate (contemporary) control group plo ior weaknesses of studies on exercise and job absenteeism. If an emyear, xercise program did reduce sick leave in the first few months of the interes ost likely would exert its beneficial effect through increased job tude) ra hanced social support, reduced boredom, improved mental attithan sudden substantial improvement in physical health.

PROGRAM NITIATION AND MAINTENANCE FACTORS

Little systematic esearch has been conducted to determine which factors influence the ini tion and maintenance of exercise programs by adults, compared to other ealth-related habits such as eating, cigarette smoking, and alcohol consumption. Many principles of behavior change learned from the study of these other habits may apply to implementation of adult exercise programs in industry, but since long-term adherence to supervised or unsupervised exercise has been a major problem for many programs, research is needed on this topic. Many of the data available on factors important for long-term adherence to exercise programs have been obtained by asking participants or dropouts about program characteristics that influenced their participation, rather than by conducting controlled trials in which selected program characteristics are systematically varied and their effects evaluated.

Heinzelmann and co-workers *(22, 23, 28)* are responsible for many of the data available on factors influencing recruitment and adherence to supervised exercise programs as a result of their studies on men in an industrial setting (NASA headquarters) and in a community setting. Some of their major findings and those of other studies conducted in industry follow.

Volunteer participation in a supervised exercise program is positively related to level of socioeconomic status. This finding is consistent with the general social learning theory that the more highly educated are more likely to participate in new health-oriented behavior. Certain health behaviors—exercise, weight control, diet—are more likely to be viewed as beneficial by persons in higher social classes than by those in lower social classes. However, persons in higher social classes may simply have more time available to participate in exercise programs, as well as more flexible schedules, than those in lower social classes who are engaged in occupations requiring a more routine and fixed schedule. Thus there is a need to establish exercise programs that allow flexible times for participation, as well as programs that can be conducted in various employment settings.

People's health attitudes and beliefs can influence their decisions to participate in an exercise program and to adhere to the program over time. Middle-aged men tend to have a positive attitude toward voluntary participation in a program because of (*a*) perceived vulnerability to disease, especially heart attack, (*b*) perceived health benefits of exercise, (*c*) feelings of control regarding their health status, and (*d*) confidence in health professionals. Efforts should be directed toward identifying and influencing the views of potential participants concerning their personal need for and the benefits of a proposed program. Special attention should be given to the attitudes and beliefs of persons whose views may not be consistent with program participation.

Discussions with small groups during recruitment can enhance the decision to participate as well as the adherence pattern. This recruitment method was systematically compared with the approach of lectures for large groups and was found superior for recruitment and adherence, independent of social or personal characteristics of the audience *(23)*. The active engagement of potential participants in group discussions is more likely to increase understanding and learning than the more passive reception of information obtained from a lecture, and it also provides an opportunity for a person to explore and evaluate the benefits and demands of program participation.

People may be motivated to exercise or participate in a physical activity for reasons other than health. Some persons may decide to participate primarily for health reasons while others may participate because the program provides a change of routine and an opportunity for recreation or social contacts, increases their fitness for other games or sports (skiing, hiking, and so on), or enhances appearance (self-image). Thus, when efforts are made to promote exercise participation, the focus should be diverse and take into account a variety of motivating factors, health-related or not.

Factors influencing a decision to participate in an exercise program frequently differ from factors that influence adherence to the program over time. Motivating factors for participation in an employee exercise program may be a concern for health, a desire for recreation, or a change in routine, while factors such as the organization and leadership of the program, types of activities offered, convenience of participation, and the camaraderie or social support that is generated may be more important in promoting program adherence over time. Since the social aspects can be significant in promoting program adherence, major efforts should be made to ensure that group exercise programs are administered in a manner that supports rather than impedes social development. Wanzel *(29)* attempted to determine why 480 salaried employees withdrew from a company exercise program. A total of 254 employees (53 percent) who had quit the program responded to a questionnaire that included reasons for withdrawal. Major reasons given for quitting included (*a*) facility too far from workplace (43 percent), (*b*) exercise program rearranged participant's schedule too much (40 percent), (*c*) usual exercise time in the facility was too crowded (14 percent), and (*d*) medical reasons or injury (18 percent). Since the exercise program was operated during nonwork hours, persons who had quit were asked if exercise sessions during office hours, two or three times a week, would have been a suitable alternative to their usual exercise time, and if this would have kept them in the program. Of the 254 respondents, 65 percent answered "yes" to the question, and 78 percent said that this type of scheduling would not decrease their normal office productivity.

Yarvote and co-workers *(30)* reported their early experience with a medically supervised exercise program for executives of Exxon Corporation. A total of 422 persons were invited to participate in the program, and 309 (73 percent) accepted. When a comparison was made between those who entered the program and those who did not, it was observed from the results of periodic health examinations that the ones who did not participate were older, smoked more, had higher blood fat levels, higher blood pressure, more heart disease, and poorer treadmill performance. Thus, at least in this instance, those who might have benefited most from the program elected not to participate. If short-term health benefits are to be obtained from exercise programs in industry, recruitment and adherence tactics need to be developed to ensure participation by the higher risk segment of the population.

According to S. N. Blair and co-workers, who surveyed 504 white-collar employees in an insurance company headquarters, several factors were apparently related to the employees' participation in regular exercise (unpublished data from a Liberty Corporation employee health study). Leisure time physical activity (LTPA) was measured by an extensive questionnaire. Factors related to amount of LTPA included age, sex, religion, locus of control, cigarette smoking, and degree of life satisfaction. When only vigorous LTPA was considered, race was also related to amount of activity, but cigarette smoking was not. Factors not related to LTPA included salary level, marital status, type A (hard driving) or type B (less time-conscious) behavior patterns, and frequency of attending religious service. Although several factors were significantly related to LTPA, they collectively accounted for a relatively minor portion of the variations in LTPA.

In a recent report of a heart disease intervention program for public employees, Bjurstrom and Alexiou *(26)* reported that 61 percent of the participants were still in the program at the end of one year, a much better experience than reported by many others. The retention rates for years two through five were 52 percent, 42 percent, 37 percent, and 25 percent, respectively; at year five, the retention rate was significantly different for men (45 percent) and women (11 percent). Of those who dropped out during the first 15 weeks, 79 percent reported doing so because of lack of interest, including lack or loss of motivation, logistical difficulties associated with the program schedule, or problems encountered with supervisors regarding time for program participation. Another 13 percent dropped out because of physical or medical problems and 8 percent because of changes in job status (transfer, retirement, or change in workload or assignment). During the remainder of the first year, 81 percent of the dropouts were reportedly due to lack of interest; the change of job status reason increased to 18 percent, and medical reasons decreased to 1 percent. Attrition after one year of program participation resulted increasingly from changes in job status and decreasingly from lack of interest and motivation, particularly among men. In addition to the decrease in the percentage of attrition associated with medically related problems, the type of medical problems that caused attrition were more often not attributable to program participation.

Attention should be given to persons who are most likely to influence a potential participant's attitudes and behavior. Often the attitudes and reactions of those with whom a person interacts (wife, co-workers, supervisors, or friends) determine whether that person will participate and adhere to the program over time. These influential persons should be adequately informed about the program and, if possible, be involved in the program on a continuing basis to ensure that their reactions provide social support and reinforcement for the participants. For example, in a study of men exercising under supervision three times a week for 18 months, the wife's attitude toward the program was highly related to her husband's adherence. Of the men whose wives had a positive attitude toward the program, 80 percent exhibited good or excellent adherence, in contrast to 40 percent of the men whose wives' attitudes were neutral or negative *(23)*. In the NASA headquarters' study, the supervisor's attitude toward the program was associated with employee adherence to the exercise program *(22)*.

SUCCESSFUL EXERCISE PROGRAM COMPONENTS

The research cited in the preceding section, behavioral research on related health habits, and the experience of numerous employee exercise programs, all indicate that certain arrangements or services are important for including long-term participation by a substantial percentage of employees.

An employee exercise program is more likely to succeed if the program is designed to fit the needs, interests, and capacities of the employee and the industrial setting. Careful planning, including development of written objectives and behavioral outcomes, is essential during the early stage. The interests of management and employees in various types of activities should be determined before a program is initiated. The number of employees, their age, sex, educational level, and current health status should be considered, as well as certain job characteristics, such as type of occupation, amount of physical activity, and travel schedules.

Following are some components that lead to a successful exercise program. No one program must have all the components, but the more of them that can be appropriately incorporated, the more likely success will be achieved.

Knowledgeable Leadership

Regardless of the type of program selected, the most important determinant of success will be the ability of the program director or leaders to properly inform and help the employees motivate themselves. Good leadership is more important than elaborate facilities or expensive equipment. Without effective leadership the chances that many previously inactive employees will continue any type of program are greatly reduced.

Program Promotion

An effective employee education campaign increases the likelihood of program success. The employees need to understand why the employer is offering the program. They need to know the potential benefits of regular participation—if hazards exist, how they proceed to become involved, and what is expected of them in terms of time, equipment, or fees. An education and promotion campaign should be continuous or repeated frequently to achieve maximal effectiveness.

Evaluation

For long-term success, active medical participation is highly desirable. Ideally, an exercise program should be an integral part of the employee health program. Minimal medical approval can be provided by the employee's personal physician. However, several problems exist with this approach: Frequently an employee does not have a personal or family physician; a visit to a physician is an added expense to the employee; and the examination may not be the type needed for clearance procedure. The best approach is for the medical personnel who provide occupational medical services to the employees to conduct any necessary clearance procedures and coordinate their results with the personal physician.

Some form of objective evaluation is required if the program participants are to know how they are progressing. The specific type of evaluation depends on the purpose of the program, but it might include body weight; resting, exercise, or recovery heart rate; simple performance tests; or a more sophisticated medical evaluation. The frequency of these periodic evaluations depends on the time and expense required to administer them, the number of participants in the program, and use of a particular test to determine how rapidly a participant should be expected to demonstrate favorable changes.

Convenience

Exercise programs sponsored by employers can be more successful than many programs sponsored by community agencies, primarily because of their convenience for the participants. A program located at a worksite that includes a shower facility and towel service, as well as a supervised exercise area, can eliminate many of the reasons adults have for not exercising regularly. They do not have to locate a facility, drive somewhere else, find a parking place, or be concerned about transporting equipment and clothing. All aspects of an employee exercise program should be geared toward convenience and safety so as to obtain optimal participation.

Activity Variety

If a variety of physical activities are offered to persons who are considering participation in the fitness program, a much higher recruitment rate will be achieved than if only a single activity is offered. Once in the program, a participant should be able to switch activities to maintain his interest as long as the new activity will fulfill the objectives of his program. Seasonal changes in climate should be considered in program variation.

Recognition for Participation

An awards program for regular participation or achievement may be an inexpensive but effective means of motivating people to continue participation. Awards should be given for various levels of achievement or participation, with differences in capacity—sex and age—taken into consideration. The giving of such awards can be based on periodic testing programs or on the documentation of regular performance (number of exercise sessions attended, distance walked, jogged, ran, swam, and so forth).

IMPLEMENTATION IN INDUSTRIAL SETTINGS

Numerous employer-sponsored exercise programs have been initiated in the United States during the past two decades. Some have been highly successful, as determined by participation, subjective evaluation by management and participants, and enhancement of physical fitness. Other programs have been dismal failures. What can be learned from these experiences that will enhance the likelihood of success for current or future programs?

Type of Program

A variety of organizational plans have been exhibited by employee exercise programs. Financial support, facilities, time, and personnel have been provided in varying amounts by management, labor unions, and such employee groups as recreation associations. In certain occupational settings, successful programs can be conducted with relatively minimal management involvement; under other circumstances, a much greater commitment by the employer is needed to ensure success.

The minimal level of effort and least expensive program could be one of exercise promotion or education. The objective of this type of program is to encourage increased physical activity by using written materials, movies, and speakers to heighten awareness and to provide information on how to exercise. Effectively conducted, this approach can enhance many employees' understanding of the value of exercise. Materials such as those provided by the President's Council on Physical Fitness and Sports, Corn Products Company International, and Continental Insurance can help employees begin a program. This type of minimum program probably will not significantly improve the exercise habits of many sedentary employees unless it is supplemented by a more comprehensive effort.

Many possible options exist regarding the organization of company-sponsored programs. Some of the major organizational components to be considered in program planning are:

Supervision—No supervision, company provides only facilities or time, or both. Consultant at rented facility, YMCA, or school. Company provides trained exercise leader.

Facilities—A major expense, if not already available.

Alternatives—Rent YMCA, YWCA, school, club facilities or unused space such as parking garage, storage room, and so forth.

Time—Program offered on company time, employee time, or shared time commitment.

These organizational components can be put together in several different ways, according to the employer's objectives and resources. The level of company commitment can range from providing minimal facilities and requiring employees to exercise on their own time to supporting a total program of trained leadership, company time, and complete facilities.

Skilled Leadership

We believe that highly trained leadership is the factor that the company should support most strongly. The exercise leader should be well trained in individualized exercise prescription, safety factors and first aid, other aspects of the overall health program, and behavioral techniques to help motivate and maintain participation in the program. Not all persons who purport to be exercise leaders have satisfactory levels of these skills, and care must be taken to obtain the best-qualified person. The American College of Sports Medicine (ACSM) has a certification program for exercise program personnel and can provide assistance in exercise leader training and recruitment. The right exercise leader can conduct an excellent program with minimal facilities, but without leadership, the most palatial facility will not produce many long-term changes in sedentary persons.

Facilities

The most important facility that the company can provide is probably a shower and change room. Aerobic exercise can be conducted in numerous places, including stairwells and hallways. However, it would be better to provide an exercise room. Outdoor jogging or walking trails can be relatively inexpensive for many companies. The ideal provision would obviously be a complete gymnasium with equipment and special exercise rooms, such as squash courts. It is strongly advised that a competent consultant be retained before spending large sums of money on facilities. An experienced ACSM-certified program director would be a good choice. Many companies have wasted money on useless, dangerous, or inappropriate exercise equipment; this mistake can be avoided by obtaining appropriate advice.

Time

A third consideration in organizing a program is when to conduct the exercise sessions. This is not a problem for managerial level employees, who generally have the flexibility to plan their schedules. Many employees who are mostly sedentary and need the program do not have the luxury. Should the program be offered to these persons on their own time, before and after work or during the lunch hour, or should they take time from their work three times a week to exercise? We favor a shared approach, which obtains a commitment from both employer and employee. There are sound behavioral reasons for this method, and it appears to be feasible in many settings.

Costs

In most instances, the decision to have a program and the type of program depend on the available financial support. Consideration should be given to the amount and potential resources for initiation of a program, as well as for its continued operation (supervision, maintenance, and so forth). Financial support from several sources can be combined. Support might be obtained from a special appropriation made by management, the employee health or medical program, labor unions, the employee recreation association, and membership fees charged for participation in the program.

As stated previously, we consider trained, knowledgeable leadership to be the key to a successful program. Salary for a full-time trained exercise specialist with a master's degree and several years' experience probably will be $15,000 to $22,000. A part-time specialist may be hired by smaller organizations, or several companies could combine their resources to hire someone to direct a jointly sponsored program (two or more organizations in a building or industrial park). Program costs beyond a program director will depend on the type of program implemented; however, the program can include additional personnel, facilities, educational material, and medical evaluations, as determined by the program design. Since in no study have the economic benefits of employee exercise programs been adequately evaluated, the cost-effectiveness of various approaches cannot be determined.

RECOMMENDATIONS

1. Exercise programs in industrial settings should be promoted by employee groups, as well as by management. They should be considered an integral part of an employee health benefit program. Their incorporation into basic health services should increase their effectiveness and substantially lower the cost per participant.

2. Support should be provided to public and private agencies to develop guidelines and manuals of operations for employer-employee exercise programs. Included in such resources are the President's Council on Physical Fitness and Sports, the American Association of Fitness Directors in Business and Industry, the American College of Sports Medicine, and the Exercise Committees of the American Heart Association and American Medical Association

3. Special attention should be given to employees of small businesses that cannot afford to provide the exercise programs or services that are economically feasible for larger business. Guidelines should be established and implemented for the development of contract services to be provided to management and employees of small businesses by local community agencies (YMCAs, YMHAs, schools, and others) or by private enterprise.

4. The major need at this time is more research on various aspects of employee fitness programs. As we have indicated in this report, few objective data exist for judging the effectiveness of existing programs.

Future planning will remain speculative unless this deficiency is remedied. Therefore, we make the following specific recommendations: Sources of funding for a greatly expanded research effort should be identified. One possible approach would be to establish a consortium, with government, labor unions, and private industry contributing to a funding pool. These research funds would be disbursed on a competitive basis to support applied research. Requests for proposals (RFP) could be developed by a consultant committee of scientists working in the area. A peer review system similar to that used by the National Institutes of Health and the National Science Foundation should be established to evaluate submissions to RFPs and investigator-initiated proposals. We recommend that these funds be used to support research in the broader aspects of health enhancement, in which exercise programs are integrated with other health promotion efforts. An initial goal of $50 million is suggested to support this type of research.

Although the list of potential important research topics on employee exercise programs is extensive, some major issues are:

- Development of strategies to motivate the initiation of, and long-term participation in, exercise programs. Special attention should be given to groups that have been slow to adopt sound health behaviors—for example, lower socioeconomic-status workers.
- The determination of the type of program organization that is most feasible. Should employees be given time off for exercise? Are group or individualized approaches most effective?
- Extensive cost-benefit analyses. Can exercise programs be effective in meeting stated objectives? Can self-help approaches be effective (thereby reducing costs)? How can exercise programs be best integrated with other health promotion efforts?

SUMMARY

Sufficient data are available that support the health benefits resulting from a physically active life-style (especially with accompanying other health-related behavior) to justify the inclusion of exercise as one component of health promotion programs in industrial settings. Exercise programs should be centered on the performance of large-muscle dynamic (aerobic) exercises, because these have the greatest potential for increasing physical working capacity and the prevention or amelioration of atherosclerotic vascular disease, obesity, and selected metabolic disorders. Attention also should be given to strength development and flexibility exercises for the prevention of orthopedic injury and the maintenance of lean body mass.

The health benefits of exercise appear to be related to the increase in the intensity (relative to a person's capacity) and amount (duration times frequency) of exercise, and thus more vigorous forms of exercise seem to result in the greatest improvement in both physical working capacity and health. Performance of vigorous exercise by adults has some cardiovascular or orthopedic risks, but these risks are acceptably small if the participants initiate an exercise program following appropriate guidelines as to medical clearance and exercise performance (guidelines available from the American Heart Association, American College of Sports Medicine, and American Medical Association).

Factors influencing the success of recruiting and maintaining sedentary employees in a more active life-style include (*a*) the involvement of knowledgeable and enthusiastic leadership, (*b*) a program in which participation is reasonably convenient, (*c*) adequate instruction on why and how to exercise, (*d*) provisions for a variety of appropriate activities to meet different needs and interests, (*e*) support by peers, supervisors, and family for continued participation, and (*f*) establishment of short- and long-range goals, with periodic assessment and appropriate recognition or awards.

Many employee physical fitness programs currently exist, and the number is increasing rapidly. The broad objectives of these programs are (*a*) to save money by reducing employee illness and to increase productivity and (*b*) to provide a health enhancement service for employees. Currently available data, though meager, suggest that these objectives are being partially met.

The specific exercise program selected for implementation will be successful if it is designed to meet the characteristics of a particular industrial setting. The age, sex, interests, and needs of the employees should be considered, as well as the industrial setting, finances available, and other health-related services. A standard program or set of programs will not meet the needs of the various industries; so expert consultation is required if successful long-term programs are to be implemented. With careful planning and the support of the business community, labor unions, employees themselves, and the scientific community, physical fitness programs can be improved and expanded to many more U.S. workers.

REFERENCES

1. *Physical Fitness Research Digest.* President's Council on Physical Fitness and Sports. Series 4, No. 2, April 1974, pp. 1–27.

2. *"E" is for exercise.* Dallas: American Heart Association, 1978, p. 6
3. *Exercise and health: A point of view.* Chicago. American Medical Association, 1968, p. 12.
4. Exercise and participation in sports among persons 20 years of age and over: United States, 1975. *Advance Data* 19:1–11 March 15, 1978 (National Center for Health Statistics, Hyattsville, Md.).
5. Fox, S. M.; Naughton, J. P.; and Haskell, W. L. Physical activity and the prevention of coronary heart disease. *Annals of Clinical Research* 3:404–432 (1973).
6. Froelicher, V. F. The effects of chronic exercise on the heart and on coronary atherosclerotic heart disease: A literature review. In *Cardiovascular clinic,* edited by A. Brest. Philadelphia: F. A. Davis Co., 1976.
7. Paffenbarger, R. S., et al. Energy expenditure, cigarette smoking and blood pressure level as related to death from specific diseases. *American Journal of Epidemiology* 108:12–18 (1978).
8. Shapiro, S., et al. Incidence of coronary heart disease in a population insured for medical care (HIP). *American Journal of Public Health* 59:1–101 (1969).
9. Wood, P. D., et al. Plasma lipoproteins in male and female runners. *Annals of the New York Academy of Science* 301:748–763 (1977).
10. Selvester, R.; Champ, J.; and Sanmarco, M. Effects of exercise training on progression of documentary coronary atherosclerosis in men. In The marathon: Physiological, medical, epidemiological, and psychological studies. *Annals of the New York Academy of Science* 301:495–508 (1977).
11. Cooksey, J. D., et al. Exercise training and plasma catecholamines in patients with ischemic heart disease. *American Journal of Cardiology* 42:372–376 (1978).
12. Hartley, L. H.: Physical training in sedentary middle-aged and older men. *Scandinavian Journal of Clinical Laboratory Investigation* 24:335–344 (1969).
13. Blackburn, H., et al. Premature ventricular complexes induced by stress testing; their frequency and response to physical conditioning. *American Journal of Cardiology* 31:441–449 (1973).
14. Ferguson, R. J., et al. Effect of physical training on treadmill exercise capacity, collateral circulation and progression of coronary disease. *American Journal of Cardiology* 34:764–769 (1974).
15. McPherson, B. D., et al. Psychological effects of an exercise program for post-infarct and normal adult men. *Journal of Sports Medicine and Physical Fitness* 7:61–66 (1967).
16. Backelund, F. Exercise deprivation. *Archives of General Psychiatry* 22:365–369 (1970).
17. DeVries, H. A. Immediate and long-term effects of exercise upon resting muscle action potential level. *Journal of Sports Medicine* 8:1–11 (1968).
18. Folkins, C. H.; Lynch, S.; and Gardner, M. M. Psychological fitness as a function of physical fitness. *Archives of Physical Medicine and Rehabilitation* 53:503–508 (1972).
19. Ismail, A. H., and Young, P. J. The effect of chronic exercise on the personality of middle-aged men by univariate or multivariate approaches. *Journal of Human Ergology* 2:45–57 (1973).
20. Howard, J. H.; Rechnitzer, P. A.; and Cunningham, D. A. Coping with job

tension: Effective and ineffective methods. *Public Personnel Management* 6:317–326 (1975).
21. Gendel, E. S. Pregnancy, fitness and sports. *Journal of the American Medical Association* 201:751–754 (1967).
22. Durbeck, D. C., et al. The National Aeronautics and Space Administration–U.S. Public Health Service Health Evaluation and Enhancement Program. *American Journal of Cardiology* 30:784–790 (1972).
23. Heinzelmann, F., and Bagley, R. W. Response to physical activity programs and their effects on health behavior. *Public Health Reports* 85:905–911 (1970).
24. Laporte, W. The influence of a gymnastic pause upon recovery following post office work. *Ergonomics* 9:501–506 (1966).
25. Petrushevskii, I. Increase in work proficiency of operators by means of physical training. *Voprosy Psikhologii* no. 2:57–67 (1966).
26. Bjurstrom, L. A., and Alexiou, N. G. A program of heart disease intervention for public employees. *Journal of Occupational Medicine* 20:521–531 (1978).
27. Linden, V. Absence from work and physical fitness. *British Journal of Industrial Medicine* 26:47–53 (1969).
28. Heinzelmann, F. Social and psychological factors that influence the effectiveness of exercise programs. In *Exercise testing and exercise training in coronary heart disease,* edited by J. P. Naughton and H. Hellerstein. New York: Academic Press, 1973.
29. Wanzel, R. S. Factors related to withdrawal from an employee fitness program. Paper presented at the American Alliance for Health, Physical Education, and Recreation convention, Seattle, Wash., 1977.
30. Yarvote, P. M., et al. Organizations and evaluation of a physical fitness program in industry. *Journal of Occupational Medicine* 16:589–598 (1974).

31

Self-Protective Measures Against Workplace Hazards

Alexander Cohen, Ph.D., Michael J. Smith, Ph.D., and W. Kent Anger, Ph.D.

[*Ed. note:* Since the intent of this book is to discuss the promotion of life-style changes that can lead to better health, there has been no stress on occupational hazards. But this article has been included because it deals so clearly with behavior change factors and techniques. Many of the educational techniques described here are relevant to both life-style change and health hazard education.]

INTRODUCTION

One may spend a major part of daily living in a work environment containing a host of real or potential hazards in the form of toxic chemicals, harmful physical agents (noise, heat, vibration, radiation), and dangerous electromechanical equipment. Job demands and psychosocial aspects of the work situation may also be sufficiently stress-producing to impose added health and safety risks on workers *(1)*.

This article was first published in the *Journal of Safety Research* (Fall 1979) and appears here (in slightly revised form) by courtesy of the National Safety Council.

The authors are with the Behavioral and Motivational Factors Branch, Division of Biomedical and Behavioral Science, National Institute for Occupational Safety and Health, Cincinnati. Dr. Cohen is Chief of the branch, Dr. Smith is Chief of the Motivation and Stress Research Section, and Dr. Anger is Chief of the Behavioral Studies Section.

Although the early part of the twentieth century witnessed significant advances in the United States in eliminating workplace safety hazards, the number of worker injuries and deaths still remained at unacceptable levels. Moreover, the prevalence of job-related diseases, as distinct from traumatic work injuries, could never be adequately assessed because of their insidiousness and latency in development, coupled with much confounding by nonoccupational influences. The enormity of the occupational disease problem is only now beginning to surface in light of expanded surveillance, epidemiological studies, and other research directed to workplace health hazards *(2)*.

This last type of activity is one result from the passage of the Occupational Safety and Health Act (OSHact) of 1970. This legislation recognized the need for a more concerted effort to reduce the human toll of work-related injury, disease, and death. The goal of the OSHact, as written into its introductory section, is to assure so far as possible a safe and healthful workplace for every American male and female worker. Consistent with this goal, the OSHact calls for the development and establishment of new and revised standards to better safeguard the workers' health and well-being, including the prescription of suitable control measures and their enforcement. In implementing hazard control, emphasis has been placed on engineering or physical schemes. Yet there is acknowledgment that behavioral, motivational, and other human factor considerations can also be important to any overall effort at minimizing job health and safety risks. Some instances where human factor considerations may play crucial roles are as follows:

1. Acceptance and proper use of personal protective equipment as an interim means for reducing apparent work hazards until engineering controls are in place, or to cover supplementary or incidental control needs (e.g., during maintenance or emergency shut-downs of main control systems)

2. Adherence to specified work practices that reduce contact with health-endangering substances and limit opportunities for accidental injury

3. Carrying out of prearranged actions during an emergency at work (e.g., spill, fire, machine malfunction, failure of engineering system for environmental hazard control) so as to avoid harm to oneself or co-workers and to contain or correct the source of the mishap

4. Establishment of a heightened awareness and understanding of job health and safety risks so as to amplify one's protective posture against assorted work hazards, even those as yet unrecognized

This paper will examine techniques for influencing individual worker behaviors, actions, and attitudes in the above directions, thus improving self-protection against work related hazards. Included here will be reported experiences with these methods from field tests and case studies. Based upon current knowledge and thinking, those approaches deemed most effective in promoting and maintaining self-protective actions among workers will be recommended for more general study and use.

FOSTERING SELF-PROTECTION THROUGH DIRECTIVE APPROACHES

Techniques exist for eliciting, shaping, and maintaining specified target behaviors in concurrence with safe work practices or other actions affording greater worker protection against job hazards. Training and contingent reinforcement strategies for these purposes are discussed in this section.

Training Techniques

Training represents a universal approach for acquiring specific knowledge and facilitating its use. Research in the learning area has yielded well-developed and accepted principles *(3)*. Goldstein *(4)* has composed guidelines for hazard-control training in industry that encompass many of these principles. They include the following considerations.

1. Informing the workers as to the need for learning job procedures emphasizing safe and healthful work practices is critical to gaining their acceptance. What the subject matter will be, how it will be taught, and inviting active involvement of the workers throughout the course are also essential.
2. The actual behaviors to be learned must be clearly identified through demonstration or other means of practice that will allow maximum carryover from the training to the work situation. The most practice time should be spent on those behaviors that are weakest and need strengthening relative to safety and health needs.
3. Approaches to trainee motivation include (*a*) making rewards explicit and contingent on the display of the sought-after safe or healthful behaviors, (*b*) establishing performance goals encompassing those acts, and (*c*) presenting knowledge of results or feedback. Suitable rewards will be treated in the discussion of contingent reinforcement below. With regard to feedback, it should be designed not only to inform the trainee if he is right or wrong but also to offer constructive criticism or added information to aid understanding.

4. Some special learning principles can be applied to safety training. Among these is the use of overlearning for controlling behavior during emergencies. Repeated, frequent drills in response to emergency conditions far beyond the point of mastery ensures that workers will remain unaffected in times of stress. Another technique is to assign trainees to worker groups exhibiting good safety practices, thus providing added social support for developing similar habits. Individual differences in learning ability must also be realized, and programmed instruction or one-on-one training when necessary.

Although the safety training literature contains numerous descriptions of new and varied program materials, there has been little reported evaluation of the merits of such offerings with regard to retention, transferability to the job, and attainment of specific goals *(5)*. Recently, two field studies were completed that included training in acquiring safe and healthful job behaviors with suitable assessment of their effects.

One of these studies was a pilot effort aimed at demonstrating the usefulness of behavioral approaches in controlling the risk of occupational cancer *(6)*. This preliminary work took place in a plant manufacturing reinforced-plastics products. Selected workers, engaged in spraying operations, were trained to follow certain work practices likely to reduce their exposure to styrene—a hardening agent found in the spray material and a suspected carcinogen *(7)*. The subject work practices involved using appropriate personal protection (e.g., wearing respirators), avoiding high styrene exposure areas (e.g., staying out of the spray booth except when spraying or setting up materials for spraying), and taking better advantage of existing engineering controls (e.g., activating floor fans and directing toward exhaust ventilation when rolling out previously sprayed materials, spraying toward exhaust ventilation in booth).

Training was administered individually to workers during one morning. Each worker was informed as to how each of the recommended work practices could reduce his exposure to styrene. The trainer stayed with the worker as actual work began, demonstrating the proper acts and correcting those not properly done. After the initial training, the instructor visited the worker at unannounced times once or twice each day to provide encouragement or feedback in observing the work procedures in question. Observers, from remote vantage points, logged the number of times that these behaviors occurred for the selected workers both before and after the training. In addition, before-after breathing zone samples for styrene concentrations were collected on these workers and urine analyzed for mandelic acid, a measure of styrene body burden.

Posttraining data indicated a sharp increase in the observed frequency of the prescribed healthful behaviors that resulted in a 36 to 57 percent reduction in personal exposure levels to styrene. Staggering the training schedule for the workers in the study verified that the observed behavioral changes were associated with the training and not with other factors. The authors noted that none of the work practices were particularly complex, and that the training seemed to be a matter of prompting the workers to engage in such behavior. Once the acts occurred, there was little difficulty in maintaining them (at least for the three-week period of observation).

A second field study took place in a large bakery complex and dealt with a training program for reducing excessive on-the-job injuries (8). Employee injury reports and job safety analyses were used to define safe and unsafe behaviors connected with different work tasks. Employee training consisted of one 30-minute session in which this information was presented via discussion and slides, with subsequent attention being given to certain safe work practices that were found lacking in the worker group. At the conclusion of the training session, the workers were shown a graph indicating a 70 percent incidence of these and other job behaviors as evidence of their current safety performance level. The employees were encouraged to increase their safety performance level and to set a goal. A 90 percent target was agreed upon. Thereafter, observations, made in the course of unannounced daily visits to the workplace by different raters, were posted three or four times a week as a means of providing feedback. These efforts resulted in increased safe behavior incidents to the 98 percent level. This level was maintained for the 25-week period of the study. The posting of such data was then withdrawn and a reversal or drop in safety level occurred. Management, taking note of this fact quickly reinstated this practice and proceeded to maintain it. For the year, the injury frequency rate of this company dropped from 35.2 to 10 disabling injuries per 1,000,000 man-hours worked.

These field training studies highlight two important points worthy of comment. One is that the training accents the positive—that is, the recognition and strengthening of desirable behaviors rather than the more negative approach stressing the unlearning of undesirable acts. A second is the potency of feedback in inducing the appropriate behavioral response. The fact that feedback is not costly relative to other reinforcers adds to its utility.

Contingent Reinforcement Strategies

The central theme of contingent reinforcement is that behavior is under the control of reinforcers (9). In other words, behavior is determined by its consequences, and specified acts may be increased in their likelihood by providing rewards whenever they occur, while unwanted ones may be made less frequent by consequent punishments. Logically, the occurrence of work accidents and job-induced disease should serve as effective suppressors of unsafe or unhealthful behaviors. Yet, such behaviors persist. The apparent paradox may be explained by noting that an accident or disease does not follow each occurrence of an unsafe or unhealthful act. Moreover, with insidious types of disorders, the discomfort incident to safe behaviors (e.g., wearing protective equipment) may outweigh the distant and improbable event of the disease itself. Under these circumstances, protective actions could be reinforced in a negative way.

Important contingent reinforcement principles are:

1. For maximum effectiveness, reinforcers should be given immediately after the target behavior occurs and not be delayed.
2. For establishing specified behaviors, there should be regular pairings with reinforcers; partial reinforcement can then serve to sustain such behavior.
3. Emphasizing desired behaviors with positive reinforcements is preferable to using punishments for acts to be avoided, since punishments can elicit a number of undesirable side effects.
4. Desired behavior requiring great effort will necessitate equally great rewards.

Contingent reinforcement approaches to solving human behavioral problems have been used primarily in clinical settings. Their adaptation to safety and health programs in industry is now being advocated *(10, 11)*.

One possible deterrent to worksite applications is that the supervisor's role may have to shift from that of an administrator and supplier of materials to that of an observer and delivery agent of reinforcers. That is, the supervisor will have to attend carefully to what his workers are doing and reward appropriate behaviors. The question of types of rewards also needs to be addressed. Employees have a great variety of positive reinforcers: bonuses, promotions, informational feedback, social recognition, special privileges (time off, preferred parking slot) whose feasibility and effectiveness will have to be considered. Recent first attempts to promote safe and healthful worker behaviors via administration of some of these rewards on a contingent reinforcement basis are described below.

Smith and colleagues *(12)* demonstrated how contingent reinforcement in the form of social praise and recognition served to increase the frequency with which shipyard workers wore safety glasses as required for specific job tasks. The shipyard employed over 2,000 shipfitters, whose welding and grinding jobs produced sparks and cinders necessitating eye protection. Accident data showed that nearly 60 percent of dispensary visits by shipfitters were for eye injuries, and fully half of these occurred in 10 percent of the work crews. On-site observations of these crews found workers to be wearing their glasses only 50 percent of the time when engaged in the aforementioned tasks.

Supervisors whose crews had the highest eye injury rates received 10 hours of instructions in contingent reinforcement. Stress was placed on (*a*) observing workers more closely, (*b*) immediately praising those found to be wearing safety glasses, and (*c*) encouraging others to do so as their work required. These crews wore their safety glasses more often over the next six months, and their eye injury rate dropped from 11.8 to 4.3 per 100 employees. Other crews in the shipyard showed a much smaller decline in their eye injury rates over the same period, from 5.8 to 4.7 injuries per 100 employees.

In another case study, less than 20 percent of workers in a textile spinning plant were found to be wearing ear protectors although their noise exposure exceeded the permissible daily limit *(13)*. Meetings with the department managers were held in which the noise hazards to hearing were described, as were the benefits of using ear protectors and the current noise standard. Subsequently, intermittent daily checks of earplug use were made, and the percentage of employees wearing earplugs was posted on a graph. Reinforcers, including social praise, coffee, doughnuts, money, and market goods, were first given to those employees who wore their earplugs and later to whole worker groups. Within two weeks all workers were wearing their earplugs each day. No follow-up data were reported beyond the fourth week of the study.

Individual feedback was shown to function as a reinforcer in another study designed to increase use of ear protection in noisy workplaces *(14)*. Baseline observations found that less than 50 percent of the workers in one department of a metal fabrication plant wore ear protectors despite hazardous levels of noise exposure. Subsequently, groups of six workers were selected for audiometric tests, each worker's hearing being tested at the beginning and at the end of the workshift. The resultant audiograms were shown to each worker and also posted for other workers in the department to view, along with a notation as to whether or not the worker had worn earplugs on the test day. When pre- and post-exposure audiograms were compared, temporary hearing loss was evident for those workers who did not wear their earplugs, but no such losses were found for those who used them. Further, older workers who had not worn earplugs during their working lifetime were tested and audiograms of their evident hearing loss were displayed to show the long-term effects of failure to use ear protection.

After the end of this hearing test phase, ear protector use climbed steadily in the department, reaching an apparent plateau of 90 percent in the fifth month. The frequency of observed earplug use in another noisy department of the same plant, serving as a control, remained at a very low level over this same period.

The status of workers' hearing, as an immediate consequence of wearing or not wearing ear protectors, proved an effective reinforcer in this situation. Moreover, unlike other reinforcers that have only short-lived effects and thus require continuous maintenance, information on the temporary hearing loss here had lasting qualities. Identifying other forms of meaningful and durable reinforcers could facilitate the use of contingent reinforcement techniques in the industrial setting. The potential value of this technique depends greatly on the careful selection of appropriate target behaviors and rewards.

FOSTERING SELF-PROTECTION THROUGH NONDIRECTIVE APPROACHES

In contrast to techniques aimed at strengthening specific self-protective behaviors, other approaches seek to establish a generalized tendency to act in safe and healthful ways through appropriate attitude change, increased knowledge, or heightened awareness. The latter approaches do not focus on actual self-protective behaviors but rather on dispositional factors that are assumed to be related to, if not controlling, such acts. In actuality, the direction and strength of this relationship is still a matter of some conjecture.

This section will highlight promotional techniques for attitude change and increased awareness that may affect workers' self-protective behavior. Emphasis will be given to communication, incentive, and management style variables.

Communication

The Source or Communicator

The supervisor or first-line foreman is the key person in any work establishment, with regard not only to production output but also to work safety and health matters. There is, in fact, some evidence that the safety level of a work crew is directly related to the job safety attitudes of its foreman *(15)*. If the supervisor believes that job accidents are avoidable through exercising proper safeguards, the work crew will probably show a low accident rate. On the other hand, if he accepts work accidents as an inevitable part of the job, his crew will probably have a poor safety record.

The foreman can affect these outcomes because of certain key factors in persuasive communications. For example, having probably come up through the ranks, the foreman can be readily identified with key workers, and this can be an attractive helpful in gaining attitude change. Another factor is the perceived power of the foreman who can mete out rewards and punishments, depending on worker compliance with job responsibilities. The last and probably most significant factor, however, is the foreman's credibility, that is, the workers' respect for his knowledge of the work, including its evident hazards, and the extent to which they see him setting a good example in following prescribed hazard control procedures.

The foreman must be viewed as a principal agent for promoting increased self-protection at work. Nevertheless, although supervisory manuals provide some suggestions for developing positive attitudes in workers regarding safe workmanship *(16)*, few employers actually furnish special supervisory training of this or any other type. Failure to recognize this need is particularly apparent among the smaller establishments where jobs are typically more labor-intensive and subject to high health and safety risks.

The Department of Labor *(cited in 17)* is supporting pilot efforts aimed at informing worker groups exposed to carcinogens (e.g., coke oven workers) about the nature of these job hazards and appropriate precautionary measures. In these studies, selected members of the subject worker groups are used as "change agents." These workers take a seminar course that familiarizes them with the cancer dangers in their occupation and with self-protective procedures; it also gives them experience with persuasion tactics. They are expected to educate their co-workers and encourage them to comply with proper work practices. It is too early to assess the effectiveness of this approach.

Content of Communications

Much research in mass communications has been directed to evaluating the effectiveness of rational versus emotional appeals to shape attitudes and behavior. This literature can be summed up by saying that, despite a few contradictory findings, a predominantly positive relationship exists between intensity of fear arousal and amount of attitude change *(18, 5)*.

But important qualifications to this relationship bear mention, especially with regard to messages related to health and safety. It appears, for example, that high fear arousal is better for inducing subjects to cease unsafe or unhealthy acts, while low arousal is better for inducing persons to submit to procedures for early detection of dreaded diseases *(19)*. Since early diagnosis and treatment are routinely recommended means for controlling potentially serious health hazards in industry and elsewhere, this latter observation deserves underscoring.

Further, there seems to be evidence that, while high fear arousal produces high intentions of curbing would-be hazardous acts or taking preventative measures, the actual numbers of persons taking appropriate action does not differ from those subjected to less fear-provoking messages *(18)*. Positive rewards directed to behaviors themselves would seem more in order here.

Current occupational standards require that employers inform employees about the safety and health hazards associated with workplace exposures to chemicals and physical agents. Such communications present difficulties. Indeed, what kinds of information about risk factors in handling possible carcinogens or neurotoxic materials would enable a worker to make a reasonable judgment about staying at a job or leaving? How can undue anxieties be allayed in explaining that, given proper safeguards, exposure levels will be reduced to those having minimal likelihood of any adverse effect? The National Academy of Science *(16)* has drafted some guidelines for these purposes that will require trial and study. Indeed, little work has been done in the occupational field to study message content in regard to these information requirements and their use in mobilizing positive attitudes for complying with need for self-protection.

Communication Mode

The bulk of evidence suggests that the spoken word has more persuasive impact than the written one and that informal face-to-face communication is superior to any media transmission *(18)*. The greater effectiveness of face-to-face communication is considered due to its usually being two-way, eliciting greater activity from the receiver, and probably commanding more attention than would be the case for other media communication. Two-way communications appear to be more effective in gaining worker acceptance of safe job practices than one-way communications, which are regarded as more appropriate for presenting safety directives or warnings *(20)*. A one-way form of communication, namely, use of posters as instructional reminders of safe work practices, have been found effective in one field study *(21)*. Overall, however, there has been little evidence of the effectiveness of one- versus two-way communication as a persuasive tool in the safety or health field.

Receiver Factors

Group perceptions of mass communication have been shown to greatly pressure a group member into message acceptance *(18)*. This finding suggests that social support can be a critical factor in gaining worker compliance with safe and healthful work practices. Directing company safety appeals to workers' families to increase safety consciousness among workers embodies this idea, but such efforts have not always yielded consistent results *(22)*. Other specific applications to workplace safety and health issues have not been reported.

It is acknowledged that problems such as lack of understanding and nonresponsiveness to health messages by persons in low socioeconomic groups will have to be overcome, since these people populate many high-risk jobs *(23, 24, 25)*. Finn *(26)* argues cogently for introducing occupational safety and health education in the elementary and secondary school curriculum. This should exploit opportunities for reaching young people before their attitudes toward work-related injuries and illnesses become inflexible and self-destructive, and for informing them before they have selected a job about its potential for endangering their safety and health. Finn indicates how occupational health and safety education could be incorporated as parts of existing course requirements in science, health education, and mathematics and describes one school's experimentation with such a curriculum. The case for this approach in benefiting future generations of workers certainly seems to be a strong one.

Incentive Programs

The use of incentive or award programs to foster increased interest in job safety has been a controversial subject. The argument against incentives is that they are no substitute for effective safety practices, which should include proper job training, housekeeping, hazard recognition, and so on. Further, it is contended that injuries or other adversities may go unreported under incentive plans because of possible loss of an award. Others argue that as part of a comprehensive safety program, incentives may be useful in making workers more safety-conscious *(16, 27)*. There is, in fact, little evaluative evidence on this subject to draw any conclusions. Moreover, the generalizability of incentive programs to worker health as distinct from job safety matters also remains to be considered. Indeed, although reductions in injury rates may depict the benefits of a safety incentives program, gains in worker's health may not be as readily apparent. Nevertheless, the success that some companies have had with these activities deserves mention since they offer some promotional ideas for enhancing self-protection against workplace hazards.

One such firm *(cited in 27)* was a large multiplant complex in the Midwest engaged in the manufacture of specialty paper products. Using different incentives for revitalizing their safety program, this company more than met its goal of cutting its injury rate in half after one year of operation. Elements of the strategy included basing each foreman's safety performance on the number of work injuries reported for his crew. The safety scores of different foremen reporting to the same supervisor, in turn, determined the latter's safety performance score, and all such supervisor scores defined the plant manager's score. Such determinations were factored into salary increases or bonuses for these personnel. At the shop floor level, tokens redeemable for catalogue merchandise were awarded monthly to all employees in a unit that reported no lost-time injuries. Groups of employees could also donate collected tokens to a community organization, which the company would then match with an equivalent cash donation of its own. In its first year of operation, the safety incentives program involved an additional $100,000 outlay. The reduction in accidents and associated costs savings in compensation, medical expenses, and insurance premiums that occurred during the same year relative to the previous one was $228,000. So, this proved a profitable experience in all respects.

A similar result was reported for a western uranium mining company that sought to add impetus to its safety programs *(28)*. An elaborate system was devised for monthly awards of trading stamps to individual workers with no work accidents, plus bonus stamps if their work group also had none. The system prorated the amount of these awards according to the degree of hazard connected with different job operations. Losses of such awards resulted from days lost each month due to accidents involving a worker and/or his work group. Property damage accidents were also figured in such penalties. For the five years this program has been in effect, accidents involving either worker injury or equipment damage have been reduced by more than half. Compensation costs over this period have been cut from $19,000 per year to $7,500 and damage costs from $222,000 per year to $40,000. The cost of the trading stamp program to the company over this time has ranged from $9,000 to $14,000 per year, which is far offset by the savings in property accident and injury losses.

Features of these two company incentive programs that bear noting are (*a*) the awards are something of value; (*b*) payoffs are devised to recognize the desired result and penalize the undesired on both an individual and group basis; and (*c*) payoffs occur at fairly frequent intervals. Another factor is that all personnel in the organizational chain are held accountable for the desired performance results.

Management Factors

As already stressed, incentive strategies are not to be considered as effective by themselves but only as part of a total safety program. Other determinants of success in safety programming have become clearer through recent research contrasting safety practices in companies with high and low work injury experience, and analyzing factors common to companies having outstanding safety performance records *(29, 22, 17)*. The more distinguishable characteristics of successful safety programs include:

1. Management has a greater commitment to safety. This was reflected by the rank and stature of the company safety officer, regular inclusion of safety issues in plant meeting agenda, and personal inspections of

work areas by top management officials, in some instances on a near daily basis. Interviews with top management in several record-holding plants indicated that they placed the same emphasis on safety as on production and sales.

2. More open, informal interactions exists between management and workers, and there are frequent everyday contacts between workers and supervisors on both safety and other matters. These interactions provided increased opportunities for early recognition of hazards, freer exchanges of ideas in correcting such problems, and thus greater worker participation. By comparison, management contacts with workers in plants having poorer safety records were more formal and less frequent, being confined largely to safety committee or other worker-management meetings.

3. Work-force stability is greater and there is evidence of personnel practices to promote such stability. The latter included well-developed selection, job placement, and advancement procedures. Also reflective of more positive management approaches was the availability of individual counseling methods in handling problem employees, including violators of safety rules.

4. There is better housekeeping, more orderly plant operations, and more adequate environmental control of noise, heat, and dust problems.

5. Training emphasizes early indoctrination and follow-up instruction in job safety matters.

6. Features are added or variations made in conventional safety practices to enhance their effectiveness. For example, "near-misses" as well as actual accidents were investigated, and safety signs were especially tailored or designed to depict more accurately the conditions of concern.

These findings offer added empirical support for a number of considerations alluded to in the preceding discussions of directives and nondirective strategies. Though recognized for their value to accident control, the same practices or simple extensions of them appear adequate to cover job health hazards as well. These program factors should create a climate favorable to assorted preventive actions.

SUMMARY RECOMMENDATIONS

This review of approaches for fostering self-protective measures against workplace hazards suggests the following as recommendations or critical considerations:

1. Training remains the fundamental method for effecting self-protection against workplace hazards. Success depends on (*a*) positive approaches that stress the learning of safe behavior (not the avoidance of unsafe acts); (*b*) suitable conditions for practice that ensure the transferability of these learned behaviors to real settings and their resistance to stress or other interferences; and (*c*) the inclusion of means for evaluating their effectiveness in reaching specified protection goals, with frequent feedback to mark progress.

   Training is a viable option, considering industry's existing responsibility to undertake such efforts. Adjustments in current programs, notably in building in more evaluative and feedback features, may involve moderate added costs.

2. Contingent reinforcement strategies also seem to hold great potential for shaping self-protective behaviors in hazardous work environments. The procedure rests on assumptions that target self-protective acts can be clearly defined along with meaningful rewards that can be given in a frequent and consistent way. Practical problems in making the observations critical to this approach and in delivering the reinforcement with consistency may limit its applications to select work situations offering close interactions between workers and supervisors.

   Contingent reinforcement strategies would require added staff training and more time in monitoring worker performance. Additional costs may be involved if rewards are monetary in nature. Social rewards in the form of praise or recognition could offset the latter, but more material rewards may be needed to reinforce acts involving more effort.

3. Communication techniques offer a means for both "selling" and "telling" self-protection needs. Attitude changes are the goal of these techniques, and credible sources using face-to-face, two-way forms of communications with suitably fear-arousing messages seem most effective for this purpose.

   Like training, supplying job health and safety information is already a part of the employer's responsibilities. Refinements aimed at improving the effectiveness of such communications would impose little additional cost burden. Introducing occupational safety and health education in elementary and secondary schools has potential for creating desired attitudes toward prevention. Cost factors need not be excessive, given the adoption of job safety and health topics into current course teaching materials.

4. Incentive plans add interest to an established hazard control program that could enhance self-protective actions on the part of the work force. Choice of awards, equitable pay-off schemes, and performance accountability throughout the organizational chain are important factors

in the effectiveness of incentive strategies. These can involve considerable added costs that may be offset by savings in reduced accident costs. Preoccupation with or continued use of incentive programs is questionable, however, and a succession of such efforts, each requiring a bigger prize than the one before, would appear unwise. Limited use of this plan is recommended.

5. Increased self-protection can arise as a by-product of management style and leadership. Expressions of top management's commitment to worker safety and health, its interactions with supervisors and workers in this regard, plus interest in developing and conserving its personnel resources, represent factors in successful safety performance. Added burdens here would seem inconsequential—these same management actions would probably ensure greater worker productivity.

Perhaps the optimum approach for fostering self-protection involves a combination of both directive and nondirective strategies. That is, communications, incentives, and management factors can be seen as providing a climate conducive to the adoption of self-protective behaviors. Training and contingent reinforcement strategies would target those behaviors and strengthen their establishment.

REFERENCES

1. McLean, A.; Black, G.; and Colligan, M., eds. *Reducing occupational stress: Proceedings of a conference.* Cincinnati: National Institute for Occupational Safety and Health, April 1978.
2. Ashford, N. A. *Crisis in the workplace: Occupational disease and injury.* Cambridge, Mass.: MIT Press, 1976.
3. Deese, J., and Helse, S. H. *The psychology of learning.* New York: Macmillan Publishing Co., 1967.
4. Goldstein, I. L. Training. In *The human side of accident prevention,* edited by B. L. Margolis and W. H. Kroes. Springfield, Ill.: Charles C. Thomas, 1975.
5. Pfeifer, C. M., Jr., et al. *An evaluation of policy-related research on effectiveness of alternate methods to reduce occupational illness and accidents.* Columbia, Md.: Behavioral Safety Center, Westinghouse Electric Corp., 1974.
6. Conrad, R. J., et al. A strategy to validate work practices: An application to the reinforced plastics industry. (In preparation.)
7. Fairchild, E. J., ed. *Registry of the effects of chemical substances.* Washington, D.C.: Department of Health, Education, and Welfare (NIOSH), 1977.
8. Komaki, J.; Barwick, K. D.; and Scott, L. W. A behavioral approach to occupational safety: Pinpointing and reinforcing safe performance in a food manufacturing plant. *Journal of Applied Psychology* 63:434–445 (1978).

9. Skinner, B. F. *Science and human behavior.* New York: Macmillan Publishing Co., 1953.
10. Fitch, H. G.; Hermann, J.; and Hopkins, B. L. Safe and unsafe behavior and its modification. *Journal of Occupational Medicine* 18:618–622 (1976).
11. Bird, F. E., Jr., and Schlesinger, L. E. Safe-behavior reinforcement. *American Society of Safety Engineers Journal* 15:16–24 (1970).
12. Smith, M. J.; Anger, W. K.; and Uslan, S. S. Behavior modification applied to occupational safety. *Journal of Safety Research* 10:87–88 (1978).
13. Miller, L. M. *Behavior management: The new science of managing people at work.* New York: John Wiley and Sons, 1978.
14. Zohar, D.; Cohen, A.; and Azar, N. Promoting increased use of ear protectors through information feedback. *Human Factors Journal* 22:69–79 (1980).
15. Crisera, R. A.; Martin, J. P.; and Prather, K. L. *Supervisory effects on worker safety in the roofing industry.* Morgantown, W. Va.: Management Science Co., December 1977.
16. National Safety Council. *Supervisor's safety manual.* Chicago, 1972.
17. Smith, M. J.; Cohen, B.; and Schmitt, N. Current occupational health motivation research. Paper presented at the Midwest Analysis of Behavior Association Meeting, Chicago, May 1978.
18. McGuire, W. J. The nature of attitudes and attitude change. In *The handbook of social psychology,* vol. 3, edited by G. Lindzey and E. Aronson. Menlo Park, Calif.: Addison-Wesley Publishing Co., 1969.
19. Leventhal, H., and Watts, J. C. Sources of resistance to fear-arousing communications. *Journal of Personality* 34:155–175 (1966).
20. Schlesinger, L. E. Safety communications. In *Selected readings in safety,* edited by J. T. Widner. Macon, Ga.: Academy Press, 1973.
21. Laner, S., and Sell, F. G. An experiment of the effect of specially designed safety posters. *Occupational Psychology* 34:153–169 (1960).
22. Simonds, R. H., and Shafai-Sahrai, Y. Factors apparently affecting injury frequency in eleven matched pairs of companies. *Journal of Safety Research* 9:120–127 (1977).
23. Rosenstock, I. M. Why people use health services. *Milbank Memorial Fund Quarterly* 44:94–127 (1966).
24. Vincent, P. Factors influencing patient noncompliance: A theoretical approach. *Nursing Research* 29:509–516 (1971).
25. Gordis, L.; Markowitz, M.; and Lilienfeld, A. M. Why patients don't follow medical advice. *Journal of Pediatrics* 75:957–968 (1969).
26. Finn, P. Occupational safety and health education in the public schools: Rationale, goals and implementation. *Preventive Medicine* 7:527–541 (1978).
27. Czernek, A., and Clack, G. Incentives for safety. *Job Safety and Health* 1:7–11 (1973).
28. Fox, D. K. Effects of an incentive program on safety performance in open-pit mining at Utah's Shirley Basin Mine, Wyoming. Paper presented at Midwest Analysis of Behavior Association Meeting, Chicago, May 1976.
29. Cohen, A. Factors in successful occupational safety programs. *Journal of Safety Research* 9:168–178 (1977).

32

Cost-Effective Health Promotion at the Worksite?

Joseph H. Chadwick, Ph.D.

> *The workplace is receiving increasing attention as a proper site for the cost-effective delivery of health care and related health information.*
>
> B. J. Schuman, *Journal of the American Medical Association,* 1980

WHAT DO WE KNOW ABOUT THIS SUBJECT?

On the surface, it would seem that we don't know very much about the cost-effectiveness of health promotion at the worksite. We have very few data on results from programs of recent origin, the ones that represent a response to the latest surge of enthusiasm for this subject. There were only three or four recent programs for which minimally adequate data have been reported in the literature before 1981. In part this reflects a lack of time for such results to be developed, and in part it reflects other barriers to health program

Dr. Chadwick is with American Healthway Services, Menlo Park, California. He is the former Director of the Health Systems Program, SRI International, Menlo Park.

evaluation. In any event, the data available from the latest cycle of health promotion programs do not provide a suitable basis for general conclusions.

We do, however, have a good deal of background information that can shed light on this subject. We know what the trends in health costs have been, and what portion of current costs are potentially controllable. We can deduce at least the upper bound of what is likely to be achievable from health promotion at its best. We can look at past experience with health promotion at the worksite, since this subject has a long history. We cannot say exactly what the cost-effectivenss of health promotion at the worksite now is, but we can say quite a bit about what it might be. At this stage of the game, this appears to be a worthwhile endeavor.

In general, the only practical way to estimate the cost-effectiveness of health programs of this type is by means of synthesis and analysis of data from various sources. Most of the problems that health promotion programs are concerned with derive from various forms of chronic degenerative disease. In a crude sense, these problems are analogous to physiological aging. These are diseases with a natural history that spans a lifetime. No one lives long enough to evaluate the end results of programs that deal with chronic degenerative disease purely on the basis of his or her own observations of particular programs. Conclusions have to be based on the analysis and synthesis of data from epidemiological trials, behavioral studies, and complex economic models. This is a fact of life that we will have to get used to, if we want to deal rationally with problems resulting from chronic degenerative disease.

WHAT GOES UP MUST COME DOWN, OR MUST IT?

Sharply rising payroll costs for health-related benefits are often cited in proposals to put increased emphasis on health promotion at the worksite. It is certainly true that such costs have climbed very significantly over the last few decades. Between 1950 and 1980 health-related costs to employers increased by a multiple of approximately 20—see Chadwick *(1, 2)* for more details. The current total of all costs relating in some way to health (health, disability, and life insurance, worker's compensation, sick leave) is now generally in the neighborhood of 10 percent of payroll.

It is easy to point to these "staggering" increases and sound the call for action. It is also easy to make the simplistic assumption that there are potential savings that are very large because the increases have been very large. This logic tends to overstate the amount of health-related costs that is controllable at this time. To understand the true facts of the situation, we need to analyze what the above-noted increase consists of.

In the factor of 20 increase, a factor of 3 is accounted for by inflation.

This leaves a factor of approximately 7 as the increase measured in constant dollars. This is still a very large increase. But much of this increase represents a fundamental change in the way that people pay for health care and the effects of ill-health. Over the last few decades, the degree to which these contingencies are covered by insurance has increased substantially, and for a number of reasons employers have agreed to pay more and more of the insurance bills. These are fundamental social changes that are not likely to be reversed, to any substantial degree, in the forseeable future. A relatively small part of the above-noted increase can be related to inappropriate health care or excessive use of health services that are of marginal value. It is this part of the increase that is most immediately controllable. Other potentially controllable elements include the long-range benefits of reduced morbidity and mortality and the reduced insurance premiums that should accrue from improved experience ratings.

Some of this increase is controllable by the employer; some is controllable, in principle, by the third-party payers. What is controllable in a practical sense, at this time, is probably not more than 20 percent of the total, or perhaps 2 percent of payroll, at the maximum. For many firms, the amount of cost that is controllable is likely to be nearer to 1 percent of payroll. These costs are certainly not negligible. But they are also not as large as is often implied in simplistic analyses of this topic.

HEALTH PROMOTION AND HEALTH COST CONTROL

It is now generally recognized that conventional health care is not very cost-effective at the margin [see, for example, *(3)*]. That is to say that you can pay quite a bit more than the average for health care without achieving significant improvements in health; conversely you also can pay appreciably less than the average for health care without any noticeable detrimental effect on your health. Hence it appears that carefully designed schemes of utilization control can eliminate a portion of the excess costs in our present health care system. This is likely to be the main route by which health care costs can be reduced in the amounts mentioned above.

The purpose of health promotion is presumably to produce health. In order to produce health, something has to happen. This happening, even if it is only a life-style change, has a cost that by and large is not negligible in comparison to its effect. In utilization control, we stop doing something with little value in order to avoid costs. In health promotion we actually incur a cost of some sort in order to first produce health and subsequently garner the savings that flow from the presence of health.

Analysis of the type mentioned above, involving an intimate mixture of epidemiology and economics, indicates that for certain selected health risks (e.g., high blood pressure and cigarette smoking) intervention at the worksite could in principle be extremely cost-effective, in terms of the total value produced per dollar of investment. But there has to be an investment, and the results produced go to health as well as to savings. Furthermore, for many health problems related to chronic degenerative disease there is an appreciable time lag (in some cases as much as 15 or 20 years) between the intervention and the achievement of full effect of that intervention. Given average rates of turnover, the employer who subsidizes health promotion may not be the one who gains the maximum benefit as far as benefit is related to the actual production of health.

Cardiovascular (CVD) and cerebrovascular risk factors, some carcinogens, alcohol, and injury control are factors for which interventions are at present the most available and, as far as we know, the most fruitful. From analysis of what can potentially be achieved with highly efficient intervention techniques applied to CVD risk factors, I would say that the savings to a company from such health promotion programs are to be approximately $40 or $50 per employee per year under ideal conditions *(2)*. Hence the net decrease in health-related costs due to health promotion on the main risk factors now amenable to control is unlikely to exceed 0.5 percent of payroll. In summary, it appears that health promotion for cardiovascular risk alone can contribute directly to cost control, but only if it is carried out efficiently, and then only in a modest way. However, we should not overlook the fact that such programs can have appreciable indirect effects on cost control and productivity—effects not strictly related to health.

WHO BENEFITS FROM EFFECTIVE HEALTH PROMOTION?

The argument for health promotion at the worksite has to be based more on the intrinsic value of health and employee relations than on the magnitude of guaranteed savings achievable by employers. The fact that savings can be achieved at all is remarkable, since for most health services costs far outweigh dollar savings. The worksite is, in fact, an ideal setting for the outreach components of the interventions against a number of disease problems, mostly of the chronic degenerative disease type. It offers an avenue of approach that is far superior to the available alternatives. As this becomes generally recognized, we can expect the pressure and the incentives for health promotion programs at the worksite to increase.

One way to look at the cost-effectiveness of a given health service is in terms of dollar cost per year of "increased quality–adjusted life expectancy" *(4)*. Under favorable conditions it appears that this ratio for selected cardiovascular risk interventions at the worksite could be as low as $1000 or $3000 per function year at the margin (i.e., for small variations above and below typical levels of care and health). This can be compared to conventional health services that have a cost-effectiveness of between approximately $10,000 to $30,000 per function year at the margin. Obviously, the average worker can afford to buy very few years at the latter rates.

Hence, under favorable conditions, employees with the problems for which interventions are provided can in principle benefit substantially from health promotion at the worksite. The employer can also benefit, but to a degree that is appreciably smaller. The main reason for this is that much of the benefit is likely to occur after the worker has left the given employer or has left the work force altogether. Of course, if health promotion programs become standard practice, all employers would benefit together. Programs of health promotion at the worksite that are well-designed and carefully executed can more than break even, but some portion of the health benefits will inevitably flow away into the public domain. Industrywide insurance programs such as some union plans provide greater assurance of the mutual transferability of benefits from one employer to another within that industry.

There is another way in which some of the direct benefits of health promotion programs tend to flow away from the employer. This relates to the way in which cash returns are generated by improved health. To begin with, the health service itself has a number of costs. These costs may be direct (e.g., screening and health education costs) or indirect (e.g., costs of an increased level of high blood pressure care). The costs will be offset by reductions in health care costs for treatment of adverse events (heart attacks, strokes, etc.). For problems such as blood pressure control, economic analysis indicates that the reduction in treatment costs will generally not fully cover the costs of the extra intervention *(5)*. For other problems, such as smoking control, one can do a little better than break even if only health care costs are considered. The cash flows that are capable of swinging the balance into the black show up most strongly in the life insurance, disability, and the worker's compensation plans. The problem is that for all but the largest companies the rates for such plans are set on a community rather than an experience basis. Hence the health promotion program serves to enrich the insurance companies (life and disability) involved. This problem has been discussed by Warner *(6)* in the context of the design of optimal health maintenance organizations.

This important problem can be solved and certainly deserves more attention than it has received to date. What is needed is for insurance companies to offer preferred rates to groups with preferred aggregate ratings on key risk factors. Given that the risk ratings were available, this would certainly be possible, since many companies already give preferred rates to individuals with favorable risk factors (and, of course higher rates to those with unfavorable risk factors). Insurance companies are currently discussing these issues *(7)* and have offered preferred risk plans for individual subscribers who are nonsmokers and people who have taken driver education courses. But so far they have not taken concrete steps to implement preferred risk plans for groups.

The willingness of insurance companies to adjust rates in proportion to risk factor measurements could have an important side benefit. It could tend to validate the merit of health promotion programs affecting variables subject to such adjustments. As previously noted, most of the benefits of health promotion programs cannot be determined directly but rather have to be estimated from complex models of the economic and epidemiological evidence. It is unrealistic to expect senior company managers to be able to follow arguments of this nature. Insurance companies, on the other hand, run their business on the basis of such evidence and are well qualified to analyze such problems. Lower rates offered by an insurance company as an incentive for successful programs of health promotion are something that a company manager can easily understand, and he or she would contribute materially to the diffusion of this important innovation.

HEALTH PROMOTION PROGRAMS OF HISTORICAL INTEREST

This is not the first time that health promotion at the worksite has been a subject of special attention and interest. It is worthwhile to review some of the earlier applications of this general concept to see if there are lessons to be learned. Perhaps the most important health promotion idea that has been pursued at the worksite is the idea of the annual physical examination. For a long time such examinations were advocated by a majority of industrial medical directors and thought to be cost-effective. These views, however, were based on little evidence. When evaluations finally were carried out, the results indicated that an annual examination involving a broad battery of tests is in general not cost-effective for a typical population of adults. The annual physical examination is now considerably less popular than it was 10 or 20 years ago.

Multiphasic screening was the next approach proposed for health promotion at the worksite. It was suggested as a lower-cost version of the annual physical examination. For a few years there was a great flurry of interest in this modality. Some labor unions were able to have this type of screening written into their benefit contracts. However, the evaluations (which came a little sooner in this instance) again revealed that the benefits achieved could hardly justify the costs of such programs *(8)*. Multiphasic screening has since declined radically in popularity.

Nevertheless, the idea of a comprehensive periodic inventory has great appeal, and the multiphasic screening concept has in turn been replaced by a still less expensive approach: the health hazard appraisal questionnaire. No definitive evaluations of this approach to health promotion have appeared in the literature; but, given the previous experience with more powerful periodic exams, there is little reason to be optimistic about the cost-effectiveness of health hazard appraisal without significant modification of behavior.

The last ten years have seen a development that appears to be much more promising than those discussed above. Instead of approaches concerned with a wide variety of problems but only one step in the process (detection or assessment), approaches have been proposed and implemented in which all steps in the service process are coordinated for a single problem. Hypertension control is currently the most important example of this type of approach. Here, finally, is some evidence that cost-effective results can be achieved.

SOME PITFALLS REVEALED BY PAST EXPERIENCE

A review of the history of health promotion at the worksite reveals some of the more obvious pitfalls that have been encountered in the past. Perhaps most important, there has been an excessive dependence on plausibility arguments: if something sounds good, it probably is good. There has been too much promotion in health promotion at the worksite. Hereafter, we will be wise to put more emphasis on data, and to challenge any program that does not provide some basis for an evaluation over time.

There has been a strong tendency in the past to confuse detection of problems with the production of health. Evaluations have shown repeatedly that the first is not synonymous with the second. If the aim is to produce health, the program must contain, or at least coordinate, all the elements of the behavior change and/or the care services required to produce the result. Part of the problem is that there has been a tendency to overlook or at least underestimate the behavioral elements in health promotion. By and large, standard techniques of notification, referral, and information-giving are not sufficient.

Since the health factors themselves (blood pressure, smoking, etc.) seem to be simple, it has been easy to underestimate the technical and ethical difficulties involved in the development and implementation of really cost-effective interventions for these problems. In particular, most program designers do not clearly identify or attempt to solve the new issues and problems that arise in dealing with what might be referred to as "low-grade" chronic degenerative disease, where the risk levels are in the mild to moderate range. What is "low-grade," in this context, is not necessarily the disease process itself but the amount of effort that can be put into prevention on any one day and the amount of benefit that one can reasonably expect to get back from that effort. Few of the doctrines developed for our highly centralized system of acute care can be transferred without modification to the problems of health promotion at the worksite. In fact, some of the concepts of acute care will have to be reversed 180 degrees in the environment of health promotion.

ELEMENTS OF A COST-EFFECTIVE APPROACH

We have noted some of the pitfalls to be avoided, but what are the most important elements to be included in a cost-effective approach to health promotion at the worksite? Based on what has emerged from the most serious attempts to design and evaluate cost-effective health promotion programs, and from other research work that sheds light on health promotion issues, I believe the following nine elements to be among the most important.

Mini-Screening, Maxi-Intervention

Wynder and Arnold (9) have argued for an approach to health promotion that puts some emphasis on screening, but maximum emphasis on intervention after the detection of a problem, terming this "mini-screening and maxi-intervention." There is little doubt that this is the way to proceed. Logically, it follows from the simple concept that measurement does not produce health, and knowledge without behavior change does not produce health. Practically, experience shows that this approach works in a number of different situations, whereas the opposite approach (maxi-screening and mini-intervention) has been shown not to be cost-effective.

Behaviorally Oriented Methods

After the fallacy that detection and notification will produce health, the next fallacy is that adding information and awareness will accomplish the task. This question has not been studied in any depth until recently. From the results of research in the last 10 years it is quite clear that methods with appreciably more behavioral content are needed. Each health problem has its own solution requirements and the methods used must be tailored to those requirements, but in general the applicable principles of behavior modification must be brought to bear on the problem. To leave out behavior is to leave out the essence of the problem.

Self-Care Attitude, Not Paternalism

In the history of health promotion at the worksite, paternalism has played an important role. Certain firms, known for their paternalistic operating style (I am using this phrase in a positive sense) have been active and persistent in supporting health promotion programs. But recent studies indicate that there is a subtle conflict between paternalism and cost-effective health promotion. The individuals who either have or develop the strongest self-care attitudes tend to be the most successful in solving their health problems (e.g., achievement of blood pressure control). When management feels a paternal responsibility for its employees, this should be a positive factor, but ironically it can have an adverse effect if it carries over into a feeling on the part of the employee that the company has taken over responsibility for the solution of his or her health problems.

Extended Supervision

An important feature of "low-grade" chronic disease problems is that they go on and on, and the new behaviors should not be forgotten, but are easy to forget. The term "extended supervision" refers to some mechanism, extended in time, whereby what needs to be done will be reinforced and supported and not be forgotten, at least until the new behavior itself becomes a habit, which may take quite a long time. Such mechanisms have been almost entirely missing in the classic approach to health promotion problems. But in recent behaviorally oriented studies of health promotion *(10)*, extended supervision has been found to be the most powerful of the behavioral techniques that have been tried in these applications.

Ultra-Convenience

It is generally recognized that convenience can facilitate the achievement of goals in a health promotion program. However, it is not generally realized how far one has to go in this direction to achieve the full effect of convenience. This is because there has been, and still is, a strong tendency to underestimate the value of time to the individual. In much of the literature there is an implication that employees should be happy to spend appreciable amounts of time at inconvenient hours to pursue health goals. This is simply not the case. Time is valuable, and the importance of the goals is real but quite finite for any one individual. In general, ultra-convenience, that is, a high degree of convenience, will facilitate health promotion programs in a substantial way. Therefore, concepts such as "flex-time" can be very important in the health promotion context.

Automation of Clerical Functions

Health promotion programs that embody careful screening, risk assessment, follow-up, and extended supervision tend to generate a considerable amount of clerical functions. If these functions are performed on a purely manual basis, the labor costs may be prohibitive. More and more, as computers become cheaper, smaller, and more flexible, it will be practical to automate an appreciable portion of the clerical functions in health promotion. This automation will tend to increase the cost-effectiveness of health promotion at the worksite.

Technically Sophisticated Design

Since most of the elements of a health promotion program are extremely complex, they do not lend themselves to casual or superficial methods of approach. A great deal of effort needs to go into the design of health promotion programs, and the designs must reflect the best that is known in the state of the art today. This requires considerable technical sophistication on the part of the designers. It implies that fairly substantial funds will generally have to be expended in program development. On the other hand, once technically sound programs are developed, they should have fairly wide applicability, since the problems of health promotion tend to generalize to large populations.

Investment by Third Parties

As a corollary to the above, it seems likely that much of the development of service programs and the delivery of health promotion services will in the future be done by third parties. The costs and resource requirements are too great to be borne by the small- to medium-sized company. It would be extremely inefficient if each company were to try to develop all over again all the capabilities required for a health promotion program. What most companies should do is to associate themselves with some kind of a grouping of firms that can see to the implementation and evaluation of programs offered by responsible third parties. Progress will be greater for small companies in this field when this becomes a trend.

Preferred Risk Rates

An important portion of the cash benefits of health promotion comes back in the life insurance and disability plans. It is unfair and counterproductive for this gain to merely enrich the insurance companies involved. Fair-minded insurance companies will realize that they have an obligation to do something about this inequity. In any event, associations representing employers should approach associations representing insurance firms and try to define various solutions to the problem. Obviously, experience rating is not the answer, because this will apply to only a tiny fraction of the workers and the firms. What is needed is preferred rates for employee groups that have developed an improved status on recognized risk factors through their efforts in a health promotion program.

HEALTH PROMOTION AND UTILIZATION CONTROL

At first glance, health promotion and utilization control might seem to be opposed to each other. One encourages health while the other discourages health care. But on closer examination one can see the possibility that they could be complementary and even synergistic. Utilization control aims to minimize the use of health services that are not really needed. It aims to reduce costs with a minimum effect on health. Still it is in a sense restrictive. It tends to take away something that the employee previously had.

Health promotion, on the other hand, aims to increase health significantly with a minimal effect on cost. It fact, such programs can produce small positive cash flows, if the gains that they generate all come back to the point of generation (as discussed above). However, the main effect of such programs is on health. They are additive. They tend to give the employee something that was not available before. Hence, it appears that health promotion at the worksite certainly can be complementary to utilization control.

But can the two programs be synergistic, that is, can the whole effect be more than the sum of the parts? No one knows for sure, but there is some reason to believe that very carefully designed companion programs of health promotion and utilization control exhibit a synergistic effect. Both utilization control and health promotion will be enhanced as an enlightened spirit of self-care and self-responsibility is fostered among the employees of a given company. It could well be that the two elements together form a stronger basis for fostering this spirit than either element alone. If so, they will be synergistic as well as complementary.

REFERENCES

1. Chadwick, J. H. Heart disease control programs. Guidelines No. 1035, SRI International Business Intelligence Program, September 1978.
2. Chadwick, J. H. Costs and cash benefits of heart disease programs at work. A supplement to Guidelines No. 1035, SRI International Business Intelligence Program, November 1978.
3. Fuchs, V. R. *Who shall live? Health, economics, and social choice.* New York: Basic Books, 1974.
4. Weinstein, M. C. and Stason, W. B. *Hypertension: A policy perspective.* Cambridge, Mass.: Harvard University Press, 1976.
5. Stokes, J., and Carmichael, D. C. A cost-benefit analysis of model hypertension control. Paper published by the National High Blood Pressure Education Program of the National Heart and Lung Institute, National Institutes of Health, Bethesda, Maryland, May 1975.
6. Warner, K. E. Health maintenance insurance: Toward an optimal HMO. *Policy Sciences* 10:121–131 (1978).
7. Kotz, H. J., and Fielding, J. E., eds. Health, education and promotion, agenda for the eighties. Summary report of an insurance industry conference on health education and promotion, Atlanta, Georgia, March 16–18, 1980 (sponsored by the Health Insurance Association of America).
8. Collen, M. F., ed. Special issue on the status of multiphasic health testing. *Preventive Medicine* 2 (June 1973); esp. Figure 3, p. 277.
9. Wynder, E. L., and Arnold, C. B. Mini-screening and maxi-intervention. *International Journal of Epidemiology* 7(3):199–200 (1978).
10. Haynes, R. B.; Taylor, D. W.; and Sackett, D. L., eds. *Compliance in health care.* Baltimore: Johns Hopkins University Press, 1979.

Index

Please send me _____ copies of MANAGING HEALTH PROMOTION IN THE WORKPLACE by Rebecca S. Parkinson and Associates @ $19.95 ($24.95 after July 1, 1982) plus $1.50 for postage/handling. In California, add sales tax. Enclosed is my check ☐, money order ☐, or credit card number (on reverse) ☐.

NAME ______________________________

STREET ______________________________

CITY ______________________ STATE __________ ZIP __________

Please send me _____ copies of MANAGING HEALTH PROMOTION IN THE WORKPLACE by Rebecca S. Parkinson and Associates @ $19.95 ($24.95 after July 1, 1982) plus $1.50 for postage/handling. In California, add sales tax. Enclosed is my check ☐, money order ☐, or credit card number (on reverse) ☐.

NAME ______________________________

STREET ______________________________

CITY ______________________ STATE __________ ZIP __________

Please send me _____ copies of MANAGING HEALTH PROMOTION IN THE WORKPLACE by Rebecca S. Parkinson and Associates @ $19.95 ($24.95 after July 1, 1982) plus $1.50 for postage/handling. In California, add sales tax. Enclosed is my check ☐, money order ☐, or credit card number (on reverse) ☐.

NAME ______________________________

STREET ______________________________

CITY ______________________ STATE __________ ZIP __________

VISA or
MASTERCHARGE

CARD NUMBER

EXPIRATION DATE

SIGNATURE

Mail to:

Box PB
Mayfield Publishing Company
285 Hamilton Avenue
Palo Alto, CA 94301

Allow 4 to 6 weeks for delivery.

VISA or
MASTERCHARGE

CARD NUMBER

EXPIRATION DATE

SIGNATURE

Mail to:

Box PB
Mayfield Publishing Company
285 Hamilton Avenue
Palo Alto, CA 94301

Allow 4 to 6 weeks for delivery.

VISA or
MASTERCHARGE

CARD NUMBER

EXPIRATION DATE

SIGNATURE

Mail to:

Box PB
Mayfield Publishing Company
285 Hamilton Avenue
Palo Alto, CA 94301

Allow 4 to 6 weeks for delivery.